Designing the Mobile
User Experience

Designing the Mobile User Experience

Barbara Ballard, Little Springs Design, Inc., USA

John Wiley & Sons, Ltd

Other Wiley Editorial Offices

John Wiley & Sons Inc., 111 River Street, Hoboken, NJ 07030, USA

Jossey-Bass, 989 Market Street, San Francisco, CA 94103-1741, USA

Wiley-VCH Verlag GmbH, Boschstr. 12, D-69469 Weinheim, Germany

John Wiley & Sons Australia Ltd, 42 McDougall Street, Milton, Queensland 4064,
Australia

John Wiley & Sons (Asia) Pte Ltd, 2 Clementi Loop #02-01, Jin Xing Distripark, Singapore
129809

John Wiley & Sons Canada Ltd, 6045 Freemont Blvd, Mississauga, ONT, L5R 4J3 Canada

Anniversary Logo Design: Richard J. Pacifico

British Library Cataloguing in Publication Data

A catalogue record for this book is available from the British Library

ISBN 978-0-470-03361-6

Typeset in 10/12pt Sabon by Integra Software Services Pvt. Ltd, Pondicherry, India
Printed and bound in Great Britain by TJ International Ltd, Padstow, Cornwall
This book is printed on acid-free paper responsibly manufactured from sustainable
forestry in which at least two trees are planted for each one used for paper production.

Contents

Preface

Hundreds of devices. Dozens of browsers. Hundreds of implementation environments. Myriad technology choices...text messaging, voice-over-IP, Java, GPS, MMS, cameras, and more. Does the connectivity matter? CDMA, GSM, 1xRTT, CDMA-EDGE, GPRS, Wi-Fi, WiMAX, Bluetooth...

And let's not forget the users. At a desk, hiding from teachers, at a cafe, at a club. Mobile phones are used instead of lighters at concerts. People use the mobile in the bathroom.

Mobile phones are not miniature personal computers, and mobile applications should not be miniature computer applications. While product design for mobile devices is not a separate discipline from desktop computer software and web site design, it does have many differences in users, user context, technologies, distribution, and research.

The mobile space is complex, but navigable. While technologies come and go, certain key principles remain the same. 'The Carry Principle' is the observation that the mobile phone, and any related or future personal communications devices, are always with the user. This simple principle strongly influences the shape of the personal communications device market, limitations users will be experiencing, context of use, and nature of the device itself. Learn how The Carry Principle affects application design throughout this book.

Designing the Mobile User Experience is intended to provide experienced product development professionals with the knowledge and tools to be able to deliver compelling mobile and wireless applications. The text could also be used in undergraduate and graduate courses as well as any other education venue that focuses on mobile design and the mobile experience.

While many of the principles in the book will be useful to device manufacturers and mobile platform creators, it is largely targeted at the vastly larger number of people designing and developing applications to run on those devices using those platforms.

The book covers the obvious – devices, technologies, and users in the mobile environment – but goes further. Included is a discussion of design patterns in the mobile space, including handling rendering differences, in Chapter 6. Chapter 5 covers general mobile design principles and sources of more specific design recommendations. Media generation for mobile is covered in Chapter 7. Research variations for mobile users are covered in Chapter 9.

Chapter 8 covers the various players in the mobile value chain, and their history, different goals, and typical decisions. Your organization will likely be in or closely related to one of these categories, and understanding what players in the other categories are doing will help decision making. Several application developers enter the mobile space thinking that a web site and some viral marketing will get their application on devices, but historically this has failed. Learn who needs to be part of your consideration.

Finally, Chapter 10 discusses an example application, from concept to design and project management. A few appendices help navigate topics like mobile markup languages, mobile domain names, capturing images for mobile display, and SMS campaign best practices. Also find a list of companies important in the mobile field and their web addresses, and an extensive glossary of mobile terms.

I owe gratitude to my entire family and network of friends for the ongoing support I have received in the creation of this book, especially with a new baby in the house. My husband in particular has had his patience sorely tested, and he has continued to support me.

Mark Wickersham and especially Elizabeth Leggett have helped with editing throughout the book. Mark is my technology go-to man, and Elizabeth understands users and art in a way that I simply don't. The two made the chapter on media possible and as good as it is. Additionally, Elizabeth patiently reviewed every chapter, usually more than once, and put together many of the graphics for me.

James Nyce spent several hours helping with the chapter on design principles as well as reviewing the first chapter. C. Enrique Ortiz graciously review some chapters near the project completion, while on vacation. This book is the richer for their input.

About the Author

Barbara Ballard is founder and principal of Little Springs Design, a mobile user experience consultancy founded in 2001. Clients have included carriers, device manufacturers, content companies, and industry associations, with projects including platform user experience, device UI design, style guides, and application design. Prior to 2001, she worked at the US carrier Sprint PCS on the user experience of devices, platforms, style guides, and data services.

Barbara has an MBA from the University of Kansas and a BS in industrial engineering from the University of California at Berkeley. She additionally has completed all coursework necessary for a doctorate in human factors and ergonomics from North Carolina State University, with significant work in engineering, psychology, and industrial design.

1

Introduction: Mobility is Different

A mobile phone is a Swiss Army knife. It is not a chef's knife or a buck knife. We keep wanting new features on the phone, like texting, voice memos, browsing, a camera, music, and television, because we would like these things in our pocket and the phone is already there.

And like a Swiss Army knife, the user experience of each of the features leaves quite a bit to be desired. A Swiss Army knife will not deliver the quality of cut a chef's knife will, nor will it fit in the hand quite as well as a good pocket knife.

Designing applications or web sites for mobile phones is in many ways the same as designing the best possible screwdriver or fishing rod for a Swiss Army knife. There is much that needs to be done before people will actually use the application – and people will not use the Swiss Army screwdriver in the same situations that they would use a full-sized screwdriver.

While the platform, user context, business context, device, and technologies involved in a particular mobile application may be different from similar desktop applications, the fundamental product design and development practices remain the same. The purpose of this book is to give product designers, software developers, marketers, project managers, usability professionals, graphic designers, and other product development professionals the tools they need to make the transition into the mobile arena.

This is not a book about technology or specific design recommendations. Instead, it focuses on the mobile users and their context.

Designing the Mobile User Experience   Barbara Ballard
© 2007 John Wiley & Sons, Ltd

It leans heavily on principles of human–computer interaction, usability, product development, business, and graphic design.

1.1 MOBILIZING APPLICATIONS

'I don't have a need for data services on my phone. Just give me a simple phone that has good reception and battery.'

I hear some variant of this from almost everybody to whom I talk about my work who is not actually in the mobile industry – although I grant that I do not talk to many teenagers about my work.

Focus groups show that real consumers are painfully aware that the web sites that they use not only would not work well on a mobile phone, but also would have little functionality or purpose. Most people are barely willing to read a long document or news story on a relatively comfortable full-sized monitor; it is difficult to know when or why a person would be willing to read the same story on a tiny screen. And willingness to pay for a service that provides text freely available elsewhere is even more rare.

This state of affairs, which is present in some degree in most of the world, is a result of some fundamental misunderstandings about what mobility means for customers and users. These misunderstandings cause the frequent failure of companies to create useful, relevant, enjoyable experiences.

Most mobile applications have been created as a miniaturized version of similar desktop applications. They have all the limitations of the desktop applications, all the limitations of the mobile devices, and typically some extra limitations due to the 'sacrifices' designers and developers make as they move applications from desktop to mobile device.

Some mobile applications have broken the 'miniaturize' trend and have enjoyed considerable success. While sound customization in the desktop environment is something done only by highly motivated users, phone ring tones have become a key component of the mobile user experience. FOX Network's 'American Idol' television show allowed the audience to vote via text messaging, and text messaging even in the United States has become extremely profitable.

Text messaging is very popular (and profitable), especially in Europe, and most of Japan's iMode traffic is actually similar short communications services. Sprint PCS did not have two-way text messaging in its earlier offerings but developed a web-based similar product which fast became extremely profitable despite having never been advertised.

While there are several factors that these successful examples share, the most notable thing is something they do not share: they are not simply desktop applications ported to the mobile environment. A well-designed mobile application, to be successful, cannot simply be some subset of the corresponding personal computer (PC) application, but rather an application whose features partially overlap and complement the corresponding PC application's features.

1.2 WHAT IS 'MOBILE' ANYHOW?

The definition of 'mobile' is slippery. Visit the Consumer Electronics Show's 'Mobile' section and you will see a plethora of in-automobile media players, both audio and video. A laptop computer is certainly 'mobile' but is used more like a desktop computer.

Other attempts to apply a name to the field have used 'wireless', describing how the device communicates digitally. This again is problematic as more and more desktop computers are using wireless communications, as are automobiles, thermometers, and likely refrigerators in the future.

One of the earliest books on user-centered design in the mobile environment has used the term 'handheld', which wonderfully captures the essence of the size of the devices in question, but allows television remote controls into the definition.

Mobile phones epitomize mobile devices, but the category also includes personal data assistants like Palm, delivery driver data pads, iPods, other music players, personal game players like GameBoy, book readers, video players, and so forth. *Fundamentally, 'mobile' refers to the user, and not the device or the application.*

Further, this book is about the business and practice of mobile user experience management, not design for specific platforms. If you are designing a Palm application, go see a developer guide for PalmOS. If you are designing an iPod application, go see a developer guide for that platform. There are a number of mobile web and Java development guides available. These resources are invaluable.

To get entertainment and information services to the mobile user, some sort of communications device is necessary. Most target users of applications already have a mobile phone or other mobile communications device, which they carry with them most or all of the time.

1.3   THE CARRY PRINCIPLE

Of particular importance to mobile users are a special category of devices, namely *personal communications devices*, or PCDs. These are epitomized by mobile phones and text communications devices like the BlackBerry and Sidekick. The principles of design and management found in this volume apply to PCDs. In this book, the terms 'mobile device' and 'personal communications device' are used interchangeably. A PCD is:

- *Personal.* The device generally belongs to only one person, is personally identifiable, and has a messaging address and ongoing service.
- *Communicative.* The device can send and receive messages of various forms and connect with the network in various ways.
- *Handheld.* The device is portable. It can be operated with a single hand, even if two hands or a hand and a surface are more convenient.
- *Wakable.* The device can be awakened quickly by either the user or the network.

For example, a mobile phone will receive a text message even when in its 'sleep', or standby state. Note that most computers, if they are asleep, can not communicate with the network.

This combination of features makes the service indispensable and the PCD an ever present part of the user's life. The service represents safety and social connection. Because the service is indispensable, users tend to carry the device with them all the time. This fact forms the core of understanding the mobile user experience.

The fundamental distinction between mobile-targeted design and design targeted for other platforms is *The Carry Principle*: the user typically carries the device, all the time. The Carry Principle has several implications on the device:

- *Form.* Devices are small, battery-powered, have some type of wireless connectivity, and have small keyboards and screens (if present).
- *Features.* Any information or entertainment features that might be desirable to have away from a computer or television, including television itself, will eventually get wedged onto the PCD. Devices evolve towards the Swiss Army knife model.
- *Capabilities.* The wireless connection, small size, and power constraints have made devices have slower connection speeds, slower processors, and significantly less memory than desktop computers.

- *User interface.* The small screen drives the device to a single-window user interface, so sharing information between applications is problematic.
- *Proliferation.* A personal, always-present device needs to match a user's needs, desires, and personality reasonably well. One form, one feature set, one user interface will not fit all.

The Carry Principle also has implications for the PCD users:

- *User availability.* The mobile user is more available for communications and application interaction than a computer user simply because the device is always present.
- *Sustained focus.* Because the user is doing so many things, there may not be sustainable time available for the device.
- *Social behavior.* Always-available connections has made attending meetings and dinner with friends a modified experience. Coordination across space allows both more and less social behavior.

Each of the above has implications for application design.

1.4   COMPONENTS OF A MOBILE APPLICATION

Any serious consideration of the design of software starts with a consideration of where the software will be used. Designers of web sites or applications intended for use on desktop or laptop computers tend to ask 'which operating system shall we target?', as computers are so standardized.

In reality, the *desktop environment* comprises a number of agreed-upon characteristics. All have a largish color computer screen of at least 800×600 pixels, a full keyboard, a mouse, speakers, and applications residing in windows. Connectivity may be slow (30 Kb/s) or fast (500 Mb/s or more), but it is generally there. In the US, landline network access is generally unlimited.

Further, the user of a desktop application is sitting at a desk or at least with a computer in the lap. There is a working surface, and both hands and attention are focused on the computer. Interaction with other people takes place only through the computer, not generally in person around the computer.

Devices in the *mobile environment* do not play by the same rules. This is not due to the lack of standards, but due to the highly varying needs of mobile users. The differing capabilities of low-end mobile phones, high-end smart phones, and alternative devices lead to a variable environment. Expect this situation to continue for a long time.

A mobile application consists of:

- a *PCD*, with its own use metaphor, browser, application environment, and capabilities
- a *user*, using any of a set of mobile devices, who could be riding a train, sitting in a meeting, sitting in a restaurant, walking down the street, focused on other tasks, or engrossed in the device and application
- one or more *application platforms*, which can include web browsers, application environments (such as BREW, Palm, Windows Mobile, Symbian, or Java 2 Micro Edition), messaging technologies (including email, SMS, MMS, and instant messaging), media environments (types of music and video players), and so forth, with new capabilities becoming available regularly
- one or more *output interfaces* with the world outside the mobile device, including screen, speaker, infrared, Bluetooth, local wireless (Wi-Fi), cellular wireless, unique terminal identification
- one or more *input interfaces* with the world outside the mobile device, including (limited) keypad, touchscreen, microphone, camera, RFID chip reader, global position, infrared, Bluetooth, local wireless (Wi-Fi), cellular wireless
- optionally a *server infrastructure* that complements the mobile application and adds information or functionality to the above
- *interfaces* between the application's servers and other information sources
- a *network* and the corresponding *wireless carrier* (operator), who enables some of the above technologies, connects the user to the Internet and other users, sells applications and other services, may specify permitted devices, and frequently defines what may and may not be accomplished on the network

In contrast, an application delivered to a personal computer operates in a more predictable environment. Operating systems are limited to approximately three, rather than dozens. There is one browser markup language, and though there are rendering differences between browsers, they are trivial and readily handled compared with mobile browsing. Influence of any sort of the end user's ISP is unheard of. There are definitely complexities associated with developing for the personal computer, but mobile is more complex in almost every dimension.

1.5   ABOUT THIS BOOK

This book is intended to help product design and development professionals make the transition from desktop to mobile with sophistication and understanding. It covers the obvious – devices, technologies, and users in the mobile environment – but goes further. *Chapter 2* discusses the characteristics of mobile users and how they differ from desktop users. *Chapter 3* presents a framework for understanding the range of mobile devices and how they fit into users' lives, then discusses the anatomy of the personal communications device. In *Chapter 4*, learn about various application presentation technologies and how to choose the best one for a project. *Chapter 5* covers general mobile design principles and sources of more specific design recommendations. Find sample mobile user interface design patterns in *Chapter 6*. Media generation for mobile is covered in *Chapter 7*. *Chapter 8* covers the various players in the mobile value chain, and their history, different goals, and typical decisions. *Chapter 9* discusses modifications of a user-centered design process for mobile applications, including modifications of user research techniques. *Chapter 10* discusses an example application, from concept to design and project management.

2

Mobile Users in the Wild

Consider a typical desktop – or even laptop – computer user. He is sitting quietly, perhaps with music in the background, looking only at the computer. Maybe he is in an airport lounge, with people swirling all around him, but he is still focusing on the computer. When he steps away from the machine, he is no longer connected to the network.

If a desktop user is in a busy office, interruptions likely abound. Telephones, personal visits, and general noise could be present. Email and instant messaging are major sources of interruption. Personal computers and their software should be designed to work with this social state of affairs, rather than assuming users will focus on a task until completion. Some software is.

Mobile users may hold some surprises:

- Adult women make up more of the mobile phone gaming market than do any other market segment,[1] breaking the precedent of years of teenage boy gaming dominance.
- The formula for a successful mobile phone game usually involves short attention, rather than a fully absorbing experience.
- Mobile users are quite skeptical about web sites on their phone, as anybody can clearly see that it is not the same experience as a desktop computer.

---

[1] Several sources, including the Telephia Mobile Game Report for Q1:2006 and Parks Associates' Electronic Gaming in the Digital Home (Q2:2006).

Designing the Mobile User Experience   Barbara Ballard
© 2007 John Wiley & Sons, Ltd

Despite the previous, an increasing number of users are interested in television on their phones. In 2006, use is quite low, but interest was variously reported between 11% and 30%, depending on the survey.

2.1   MOBILE USER CHARACTERISTICS

To some degree, there is no particular difference between mobile users and the users of other devices. In fact, the low cost of mobile devices relative to computers, particularly combined with the high cost of laying telephone cables to remote villages, means that the mobile phone is becoming the predominant mechanism to access information services. Thus desktop users will soon be a subset of mobile users.

All this is true, but it misses the key point of mobility: most of the mobile users are not sitting attentively at a desk or passively on a sofa. They are out and about, they are social, they are moving. They use the device for more personal purposes than a television or even a computer: it is more likely to be used by just one person.

Figure 2.1 illustrates many of the issues of mobile users. Fashion is a consideration. Size is important. The device is always present, always carried. The user is interruptible.

2.1.1   Mobile

Mobile users are mobile. They may be mobile while actually using an application, or they may move between instances of using the application. Being mobile means that user location, physical, and social context may change, that physical resources cannot be relied upon, and that physical world navigation may have to be accomplished.

The user may be in rush-hour traffic, in a meeting, in class, on a train, walking down the street, at a café, at the library, or in a restroom . . . in unlimited, ever-shifting environments. Except for highly task-focused applications, like discovering when the 56 bus will arrive at stop 70, the user's context will not be predictable. The user's context may be discoverable using current and future technologies.

Generally mobile users can be expected to have their wallet, keys, and phone, and companies are working hard at making the wallet and perhaps the keys unnecessary. What is not present is a pencil to jot down information, a user's files, reference books, or anything on the desk. Information or content stored on the computer may or may not be remotely available (typically not).

Figure 2.1 Mobile users have different availability, context and interruptibility than do desktop users

Navigating through the physical world, managing obstacles and picking routes, is a task that uses a majority of a person's attention resources. Similarly, navigating through the virtual world, performing text entry, and reading text, consumes cognitive resources. Because these tasks are similar – both navigation – they clash with each other. Typically, a user attempting both simultaneously will end up performing the tasks in sequence, or alternating. Even when alternating virtual and physical tasks quickly, either or both can suffer.

Shifting context and navigation conspire with other factors to make the mobile user more interruptible and easily distracted than desktop users.

2.1.2   Interruptible and Easily Distracted

The mobile user has all the sources of interruption from the physical world that the desktop user has, without some of the social cues that suggest he is unavailable for interruption. He is not sitting in an office, he is not facing a computer obviously focused on a task. He is instead at a client's office, at dinner, waiting for a train, in a meeting, on a date or at a desk, among many possibilities. In many of these cases, his mere presence in a public, social space could indicate he is interruptible. The smaller screen size seems to block fewer people, it is easier to meet his eyes.

He is using a device that can likely display only one thing at once, so using open windows as reminders does not come easily. Further, even the device can interrupt itself, with incoming calls or text messages. Many of his distractions cannot be stalled by social cues: the train will not wait for him to finish a task or conversation. The user therefore has no opportunity to 'just finish this sentence' when interrupted. The transition between virtual and physical tasks can be jarring and can reduce effectiveness at both tasks.

These user characteristics have a number of immediate implications for application architecture, especially in the area of state management. Most applications should, if not explicitly exited by the user, return to the same view with the same data as when the user last departed. Data should be saved without user action, possibly in a temporary store before committing changes to the official document. Because the user may not have an opportunity to save data, the application must save any critical or difficult to enter data for later reuse.

2.1.3   Available

The converse side to interruptibility is that mobile phone users are quickly available to remote friends, family, colleagues, and clients. This fact has led to higher job stress and less quiet time, but it also enables people to feel more connected.

Most personal communications devices (PCDs) are with the user constantly, either throughout the day, or throughout relevant portions of the day. These devices are likely to go with the user even to the restroom, particularly as they tend to be either worn or in pockets. Many people even feel uncomfortable when uncoupled from their devices. Thus a characteristic of mobile users is that they are present and immediately available. They are likely to look at their PCD even when they are with others.

At colleges, a large percentage of pedestrians stroll through the campus with a phone stuck to their ear, or perhaps stopping occasionally to text. No one need ever be alone. While this fosters the sense of connection to remote friends, it is also making it more difficult for people to communicate in person. A post-class conversation while walking to lunch is less likely to occur if all the students immediately dial to coordinate lunch with somebody else. Mobile phones are changing the college experience.

Culture, generation, context, and personality combine to maintain an 'importance hierarchy' for various interaction sources around the user. An in-person conversation with a respected elder is likely to trump an incoming call, but the incoming call might take precedence over a conversation with a clerk. A call from a wife or daughter nine months pregnant is likely to trump almost anything including lecturing a classroom.

Being readily available means that people answer their phones, either with voice or text, in what used to be considered inappropriate places. Texting and even voice calls in public restrooms are becoming more common. Accepting a phone call during a personal conversation has become very common, and is frequently a source of tension between different generations.

While turning off the phone, or simply not answering it, is one popular method for dealing with the phone's prolonged intrusion into life, many users do not turn it off. Ethnographic research has revealed that mobile users in Madrid think that it is rude to let a call go unanswered, and will answer it in class, when out with friends, or at the cinema.[2] Behavior differs from country to country and user to user, but even a person who does not answer the phone remains readily available. She may return the call quickly or text back, and she immediately knows the call was made.

Availability allows applications to communicate with instant messaging-like technologies with confidence that the user is present and will receive the information immediately. An application that required a return receipt from the device could ensure that a message actually made it to the device.

---

[2] Lasen, Amparo, 2002. A comparative study of mobile phone use in public places in London, Madrid, and Paris. University of Surrey Digital World Research Centre.

2.1.4   Sociable

While mobile users are available to connection from people using the phone, they are also sociable with the people around them. The other people in the restaurant are likely aware of any voice conversation, and friends at the table may be excluded from an incoming connection, or could as easily be included. A group of Japanese youth may pull out their phones to decide where to meet for dinner.

Social behaviors will vary based on who is physically present, where the presence is, the current mood, the type of incoming communication, and the source of incoming communication. An application also could be launched as part of a group activity. Consider a story:

> A small group of friends sit around a dinner table, talking about the events of the day and their friends. A phone rings. Two people reach for their pockets, and it's Larry's phone. He answers the phone, and is immediately merely 'near' people – he is 'with' the person on the phone. The conversation at the table slows to a halt, with some people starting to look uncomfortable. Conversation slowly returns once Larry is off the phone.

And a variant:

> A small group of friends sit around a dinner table, talking about the events of the day and their friends. A phone rings. Two people reach for their pockets, and it's Larry's phone. He discovers a text message from his girlfriend, and he quietly chuckles. He dashes off a response, during which time he is 'near' people. He re-enters the conversation as soon as he hits send.

And finally:

> A small group of friends sit around a dinner table, talking about the events of the day and their friends. A phone rings. Two people reach for their pockets, and it's Larry's phone. He discovers the latest installment in the mobile trivia game is available and he immediately starts the game. He reads the questions out to his table mates, soliciting opinions and gaining laughter. They decide to finish dinner and go discover the answer to the third question: 'What is the title of the book being read by the statue on the West side of the Plaza near the theater?'

In each case Larry interacts with his phone, but he interacts differently due to application technology (voice call, text message, interactive game via messaging or Java), social context, and personal and cultural proclivities.

The application type provides part of the context. A voice call is socially and technologically assumed to be between two people; adding extra parties is enough of a violation of normal behavior that announcing the presence of others in the conversation is considered a minimal requirement for politeness. Text messaging connotes a variable amount of privacy, and games have no privacy requirements.

Personal and cultural practices also provide some of the context. Larry could have deferred the call until later. He could have had his phone on silent, and made the choice based on incoming caller. He might have deferred a social call if at lunch with his boss, and accepted a call from the boss if at lunch with friends. He would have deferred the call if in a Japanese train, but might have taken the call if in a Spanish theater.

Larry is managing several 'microcontexts' simultaneously. First, his dinner companions provide a social context, both long-term and immediate. Their current topic of conversation might encourage acceptance or deferral of a call. The composition of companions and the group's history and personalities also influence call acceptance. Second, the larger physical environment – home, café, diner, or upscale restaurant – guides expectations and provides another microcontext. Third, each application – voice, text, or content – provides its own microcontext. Finally, the personalities on the other side of the mobile connection – girlfriend, boss, impersonal application – provide another set of microcontexts.

A social mobile user can manage several microcontexts simultaneously; other mobile users remove themselves from as many microcontexts as possible to focus on just one or two. Nevertheless all mobile users are exposed to one or more microcontexts. Most microcontexts, as noted above, are social microcontexts. Applications can be designed to encourage sociability in person as well as online sociability. Sociability is a key metaphor in mobile applications, and the better it is understood, the better the change of increasing application exposure and driving revenue.

2.1.5 Contextual

The mobile user's environment affects how the device is used. Ideally, the device would know whether the user is in a meeting, on a business trip, snow skiing, asleep, driving, or any other activity, and would give this information to the applications so they could behave appropriately for the user in that environment. Devices don't really do this yet, but there is a lot of information potentially available to applications that goes unused. Consider:

- A calendar application could switch the phone's ringer to vibrate and intelligently communicate to the caller that the recipient is in a meeting right now. The caller could indicate message urgency – or leave a message or call later – and the recipient could decide whether to accept the call.
- A travel companion application can use the user's location, the flight number, current flight status, and current traffic conditions to alert the user fifteen minutes before she needs to depart for the airport. The same application could alert meeting attendees when the application owner is going to be later.
- A restaurant coupon application could send coupons at lunchtime when the user is away from home and near restaurants.

Future devices may have acceleration sensors, temperature sensors, fingerprint readers, and any number of other information sources we do not currently imagine.

2.1.6 Identifiable

Because devices are personal, they are usually unique to a single user. Exceptions to this rule are rare. This identification includes both the unique messaging address (phone number or email address or similar) as well as the device.

Further, in some ways the user's messaging address is more valuable to the user than the device itself, since it is a persistent method of contacting the user. Not only is the user associated with the address, but the use of the address is directly connected to how much the user's charges will be for the month. This value is so high that special regulations in the United States mandate number portability between carriers.

In theory, subscriber identification provided by the device can be used to identify a returning user to a web site without user input. In practice, some carriers have hidden this information to all but business partners. Web applications must use cookies to identify users. However, even more than in the desktop world, there is a reliable user identification for application security: phone sharing is rare, and a missing phone is likely to be disabled so it cannot connect.

2.2 GROUPS AND TRIBES

Mobiles play a complex and evolving social role, from status symbol to facilitator of gossip.

2.2.1 Voice and Texting

Fundamentally, the mobile makes immediate long-distance relationships, to the point that long-distance relationships can become more relevant than the relationships with people nearby. The mobile combines the advantages of the landline phone, with the advantages of email, and improves upon them by being always with the user.

The idea that mobiles foster community is supported by certain research. A study by the Social Issues Research Centre, for example, looked at the role of mobiles as they facilitate gossip.[3] Gossip is used both as a connection method and as a mechanism of 'social grooming', reinforcing what is and is not acceptable behavior and hence strengthening what is and is not part of the social group. The mobile provides a constantly available mechanism to engage in immediate gossip about news, public figures, or Joe in the next office over. The mobiles enable significant social bonding: more than landline phones.

Texting adds to the social connections, but through different mechanisms and with different benefits. Teenagers can use the act of writing to be a bit less awkward in social interactions. People can send a little 'I'm thinking of you' type message to others, building the community and without the risk of a prolonged discussion or interruption. This type of interaction is beginning to replace similar practices of interaction with the neighbors to build social bonds.

While mobiles are making at least some people less interactive with their immediate surroundings and less social with people nearby, they simultaneously are having a second effect. The always-available communications reduces the risk of going somewhere alone, either through safety concerns or through group coordination challenges.

---

[3] Fox, Kate, 2001. Evolution, Alienation and Gossip: The role of mobile telecommunications in the 21st century. Oxford: Social Issues Research Centre. http://www.sirc.org/publik/gossip.shtml

This added freedom is allowing at least some people more interaction with a wider variety of environments and people than they otherwise would have experienced.

2.2.2   Extending Online Communities

Add to the simple communications properties of phones a variety of web-enabled applications that foster online communities. Myspace.com, Flickr, and various blogging sites, for example, are becoming mobile enabled. Users can get constant access to the communities, which frees them from their computers a bit as well as extends the time and degree of interaction with the services.

The process of extending an online community to mobile typically starts with adding mobile viewing capability. This step is not particularly exciting, but can serve to draw users into extended use. Use can be extended further by adding the ability to post text from the mobile, especially for sites targeted at already-texting youth.

The application can get more interesting, and more integrated into users' lives, when the camera and microphone are integrated into the application. Now users can make podcasts, provide pictures, and provide back to the community not just summaries of events, but records of events as they happen. A video clip captured at a concert, child's soccer game, or in the schoolyard can be shared on YouTube for the world – or just friends – to view. The tapestry of services available extend current online-only communities into more immediate and richer interaction, increasing the addictiveness of the services.

2.2.3   Physical and Mobile Hybrids

A new type of community-building service is developing: hybrid mobile–physical. Technologies such as near-field communications (Bluetooth, Wi-Fi) and location enable physical interaction, mediated by the mobile. The types of service provided by these communities usually have desktop access almost as an afterthought, perhaps just for signing up and configuring the service.

Geotagging, for example, is the focus of several start-up companies. The idea is that people can tag, and comment upon, a physical location in much the same way a service like Digg allows users to tag and comment upon arbitrary Internet stories. Similarly, physical world games using location tracking of other users in the game make a giant playground out of a city.

Some services enable connections to be made with people in the users' social or business networks. Some match-making services alert the user when a person with a compatible profile is nearby; other models exist. Business networking services are also available. The idea behind the business networking is simple: enable the ability to obey the oft-repeated advice, 'never eat alone'. The user consults with the service to see who in the network, perhaps a second- or third-degree connection, is nearby; a quick text message helps decide whether doing lunch is plausible and desirable. Many social dating services work similarly, but are more likely to be used in a bar than a conference hall.

2.2.4   Mobiles as Status

For most of their existence, mobile phones have provided some sort of presumed and visible status to their bearers. They started as indications of the bearer's importance or perhaps wealth. As they became smaller and less expensive, the presumption of wealth declined, but the presumption of importance remained.

Ring tones can also provide status. The default Nokia ringer is perhaps as well recognized as AOL's 'You've got mail' sound. Downloaded ringers provide enormous customization but also an indication of the user's personality. The 'mosquito' ringer, inaudible to most adults, provides teenagers the ability to differentiate themselves from adults – especially teachers.

Mobiles have had impact on the physical appearance and capacity of heavy users. Some users experience repetitive stress injuries from large amounts of texting. Many users, particularly youth, have experienced a shift in dominance of hand muscles, and their thumbs become more perpendicular to the body of the hand than their parents' thumbs. This physical shift in thumbs, and indeed the use of thumbs as the primary input method, has spawned the term 'thumb tribe' or 'thumb generation': perhaps the ultimate status symbol.

As mobiles have become smaller, they have also become fashion statements. Japanese and Korean youth wear phones on necklaces. Nokia has long provided decorated face plates. Motorola, with its RAZR and StarTAC, is good at creating fashionable devices for the tech and business crowds. Some high-end carriers promise a new phone every two months. Nokia has created the solid gold phone, for tens of thousands of dollars; they will indeed replace the innards of the phone as technology demands.

2.3 INTERNATIONAL DIFFERENCES

A common mistake is to assume that the mobile environment in one's home country is replicated internationally. This assumption is not only wrong, but it can lead to very costly mistakes. Differences include mobile industry structure, pre-existing telecommunications environment, and cultural differences. These combine to create different expectations and success conditions for applications and services.

Japan's iMode has been a major success for NTT DoCoMo, the Japanese carrier, while the competing Wireless Markup Language[4] (WML) has largely been a failure.[5] American carrier executives visited Japan to understand the business and technology and proceeded to implement similar business models on their home turf. Perhaps the biggest error was marketing these Internet-based mobile sites as 'The Internet in the palm of your hand!'. Americans, who have prolific access to computers, phones, and Internet access, did not believe that they would have a good experience on a text-only 10–20 Kb/s connection with a text-only phone. Europeans felt the same, especially since they had a successful text messaging[6] environment. This marketing error still affects how people view the mobile Internet.

On a lighter note, European bloggers have written 'how to' lists targeted at US consumers intended to encourage Americans to rely on their mobile phones more. The key recommendations include leaving the phones on all the time, carrying the phones all the time, and giving out the mobile phone number as the primary phone number – all things European mobile users do as a matter of course. These recommendations were written assuming that the calling party pays for the call – but in the US mobile phone calls are charged to the mobile phone owner regardless of whether they are incoming or outgoing. The recommendations were useless in the US environment since American

---

[4] This is commonly referred to as WAP, or Wireless Application Protocol. In this book we will refer to the markup language rather than the access protocol, to maintain consistency with the desktop Internet. After all, web sites are HTML sites, not HTTP sites.

[5] Both WML and iMode's cHTML (Compact HTML) have been superseded by XHTML Basic. Some devices have WML and cHTML extensions that thereby constitute XHTML Mobile Profile.

[6] Technically known as SMS, or Short Message Service. This is a store-and-forward text messaging service for short (usually up to 160 characters) messages.

users would not want to pay for the experience of having a telemarketer or stranger call their mobile.

Not only are the users and their contexts different, the industry itself varies significantly between regions, particularly in the relationships between carriers, device manufacturers, and content providers. In Europe, expect the device manufacturers to have the majority of the power.

2.3.1 Europe

Perhaps due to Europe's recent history developing a cross-national, consensus-based government, European industry tends to avoid jumping to market with the latest technology. Companies instead collaborate and develop standardized technologies that all companies can share. The manner in which telecommunication standards and policy are created and implemented supports this. The development of digital GSM (Global System for Mobile) in the 1980s in Northern Europe, rather than adopting analogue mobile technologies, was due to this consensus building process.

A key feature of the GSM system is the Subscriber Identity Module on the inserted smartcard, or SIM card. It stores user and billing information, including mobile operator and phone number, so that a user can theoretically use any GSM phone with a single account. Mobile operators do not have to manage phones as much as they have to manage SIM cards. Without such a card in the phone, the phone will not work.

The uniform GSM system allowed mobile phone manufacturers to create a single phone that would work for all European and other GSM carriers, instead of having to target phones at different carriers. Further, users could take a phone designed for one carrier and use it with another carrier. This meant that consumers could freely choose between devices, independent of their decision in choice of wireless operator. Further, phone manufacturers could spend engineering and design effort focusing on features rather than on carrier requirement compliance.

It also meant that the carriers were able to create near-universal, redundant, cell coverage, especially compared to American digital coverage. Thus they could compete neither on coverage nor handset selection. This advantageous environment for device manufacturers is likely what has given them most of the control in deciding what devices get designed and shipped in Europe.

Social Factors

The key factor to remember about Europe is that it is not one culture but many. Expecting a Swede to behave as a Greek will leave you with an application one group or the other may not use. British passers-by will probably pretend to not see or hear a person chatting on the phone, but the French are less likely to feign ignorance. Be sure to design for all the markets you are targeting, not just the one sharing your corporate language.

The major social factor in the development of the current mobile telecommunications environment was the existing landline telecommunications environment.

Telecommunications Environment

Much of Europe has had expensive landline phone access. Phone calls can be costly. Internet access, even in 2003, was typically found only in work environments. Phone bills might not be itemized and thus not predictable, leading some people to avoid using the phone at all. Protected monopolies eliminated any need for incumbents to change.

European operators made a pair of decisions different from US operators that have had far-reaching effects – calling party pays and cheap SMS. Whereas US operators charged the mobile users for receiving a call, European operators put the cost of mobile termination on the shoulders of the calling party. This required a separate numbering scheme for mobiles to ensure the calling party knows of the incremental charges, but encouraged users to leave the phone on and take calls.

Since the European operators were not expecting SMS to make money, they priced the service inexpensively. SMS was the cheapest way to send a message of any flavor to another person, and it was always available from the phone. Its convenience and price made the service very popular. It was powerful due to the standardization of mobile services: SMS worked across carriers. When American operators saw how popular the service was – despite the cheap American access to the Internet and email – they priced SMS at five to ten cents per message for something the user could get for free elsewhere.

Mobile telecommunications provided other advantages compared with landline telecommunications companies (telcos). Some operators itemized bills, introducing some competition. Equipment was generally less regulated. And, of course, the mobile was carried with the user.

All this has combined to make mobile penetration quite high – over 100% in Italy, Sweden, and the UK in 2004. Analysys Research expects mobile penetration for all of Western Europe to reach 100% in 2007.

Mobile Data Usage

While mobile penetration is high, data usage varies. SMS, though sometimes not considered as part of data, enjoys significant success. Web browsing and MMS has not been as popular. SMS popularity has derived from several factors:

- low cost of sending a text message compared with making a voice call or accessing the Internet via landline
- sending party pays encourages people to subscribe, since they can easily control their costs
- carrier interoperability means that users can send messages to people on any network (US carriers did not have interoperability until 2000 or so).

Web browsing, multimedia messaging services, mobile video, and similar services have not had similar success. This has been due to:

- marketing missteps – asserting that it is 'The Internet in the palm of your hand' or, more recently, 'Television on your mobile' is patently absurd because both services had significantly less choice than their full-sized counterparts and a much worse user experience with both screen size and quality
- lack of usability – difficulty in setting up a mobile for Internet access, browsers that automatically exited when connectivity dropped, browsers that then returned the user to the home page when restarting, and very difficult to use applications
- lack of consideration for mobile as having different needs – for example, replicating desktop browser behavior on mobiles such as returning to the home page upon starting the browser causing any interruption to abort the user's task

- lack of interoperability – while SMS works largely the same across devices, MMS, video, and web all have cross-device rendering issues, making it more difficult to provide content to everybody
- operator push for a 'walled garden', providing access solely to the applications approved by the operator
- lack of a compelling business model to make the creation of compelling services worthwhile.

2.3.2 Japan

After World War II, Japanese business has been dominated by clusters of trans-industry corporations with close working relationships. Each cluster is called a 'keiretsu' and typically includes at least one bank. Corporations in a keiretsu have preferential or even exclusive rights to provide services to one another.

The Japanese mobile phone is the 'ketai', and the best way to research devices, carriers, platforms, and the industry is to use that word. Mailing lists discuss ketai to the exclusion of mobile phones in other parts of the world, and the Japanese are proud of their global technology and industry leadership.

Social Factors

Japanese living conditions, especially for youth, are crowded and expensive. Landline phones are shared. Computers are shared. Youth often stay with their parents for years. Thus the ketai is the first personal (individual) method of communications a young person has.

Relationships are very important in Japanese culture, so tools that facilitate communication have a receptive market. Some iMode applications created virtual girlfriends, which would be happy, sad, demanding, or needy based on whether the user communicated with her, sent her virtual flowers, or performed other virtual relationship maintenance tasks. Thus it is no surprise that iMode's email offerings constitute the vast majority of iMode use, especially since the company does not have SMS.

The Japanese tend to have the most features on their handsets, and they tend to use them. An infoPLANT survey[7] in late 2005 found

---

[7] Translated and summarized by What Japan Thinks, at http://whatjapanthinks.com/2006/01/19/mobiles-are-alarm-clocks-cameras-and-calculators/

that, in addition to voice calls and mail, at least half of users also regularly use:

- alarm (85%) – a later survey found that 43% of Japanese users actually use this to wake up every morning
- still camera (83%)
- MIDI ring tones (82%)
- calculator (80%)
- games (66%)
- optical code reader (54%)
- high-fidelity ring tones (MP3 and similar) and videos (51%)
- calendar (51%).

Other items in the list included video cameras, remote control, music, and electronic wallet. Perhaps more interesting was the last item: 'None of the above', as only 0.3% of users selected this. That means that 99.7% of users used features beyond mail and voice, even if the services did not require connectivity.

Telecommunications Environment

The Japanese mobile industry functions as a keiretsu, to the point that Richard Meyer, in his J@pan Inc article, called it a 'keitairetsu'. Although not as formal as earlier keiretsu, it is dominated by the operator NTT DoCoMo but also includes such industry giants as NEC, Sony, and Matsushita (which includes Panasonic). The result is that DoCoMo sets the technological and service trend for the entire ketai industry.

In contrast with the European industry, the top tier ketairetsu players provide detailed device specifications and have historically developed their own standards. Lower tier players, which include foreign companies, may see the specifications after the first devices have gone to market. DoCoMo introduced iMode, for example, in 1999 using a proprietary version of HTML targeted at mobile devices.

This has started to change, with ketairetsu involvement in standards bodies such as the Open Mobile Alliance. However, the Japanese are likely to implement proposed standards long before they are formalized, making the ketai implementation vary from the standard.

This industry structure allows NTT DoCoMo, in particular, to create services that require deep handset integration. Japanese companies were the first to launch services like mobile wallet and video phone.

Mobile Data Usage

Wireless industry executives have made pilgrimages to Japan to understand why iMode and ketai in general is so popular. They have focused on the technology and on the services, a few have looked at the price. They have missed the social factors listed above, the integrated design of the handsets, and the fact that the majority of Japanese wireless data usage is messaging. The executives failed to notice the entire iMode ecosystem and have thus failed to replicate the success elsewhere.

Japanese mobile data usage is high, but only 20% of customers used their phone for more than just voice and text messaging in DoCoMo's fiscal year 2004.[8] The iMode ecosystem, with its many services, do not lure everybody into using horoscopes, shopping lists, and dating services. Mobile wallet use sits below 10% as of 2006.

As 3G handsets became more stable and less expensive, adoption is increasing. NTT DoCoMo is not making great conversion to the new services, but KDDI has a very high conversion rate. The Japanese handsets are more advanced than their European and American counterparts, and advanced features are starting to be used.

2.3.3   United States

The United States is generally considered to lag Europe, Japan, and Korea. This lagging is ascribed to a combination of ineffective companies and a less educated market. Certainly mobile phone penetration is lower, text messaging is less popular and lags European use, but this is changing. Regardless, the size and affluence of the market as well as the entrepreneurial environment mean that the country must be considered.

Social Factors

American teenagers are accustomed to having their own room, perhaps their own car, and frequently their own phone line and phone number. Computers and cheap Internet access are common, particularly among those who might use a mobile phone for data access. Local phone calls are free. Email is not quite ubiquitous, but certainly normal.

---

[8] Note that the services listed earlier were from a survey accessed by a link on the iMode home page that was present for two days, so the results are skewed towards frequent users.

GameBoys, televisions, and a constant barrage of media are typical for the American teenager.

In the 1980s and early 1990s, workers who needed always-available access, including doctors and repair technicians, generally used pagers. An entire pager shorthand evolved, with people able to send sophisticated short messages using only digits. Text pagers made this communication more robust but did not take over the market, coming late to the game. Pagers were so popular that, even when mobile phones had both text and numeric paging available, pagers remained a typical part of the worker's belt load.

The mobile phone represents yet another way to connect to others, and more expensive than either landline voice calls or email. The need just wasn't as high as it was in Japan and Europe.

Telecommunications Environment

In the 1990s, Americans enjoyed unlimited local phone calls, including dial-up access for the Internet, for a flat rate. Local access might cost around $25 per month; Internet might cost another $20. Even long-distance calls had dropped to pennies per minute. The calling party paid for calls. Teenagers spent hours chatting on the phone; computers were set to automatically redial to the Internet provider whenever the connection was dropped.

American wireless carriers selected different technologies, including analogue (AMPS), CDMA, GSM, and TDMA. They created systems that could make voice calls to each other, but that was the limit of the interoperability. Text messaging was not interoperable: in fact, many US carriers supported mobile termination only. Users could not send a text message from their phone, or if they could, it could only go to phones using the same carrier.

Paging networks had become popular. Inexpensive paging service sent a phone number only, and advanced services sent text messages. Some pages could even reply to messages, although most presumed the message would be returned by a voice call. Pagers had become integrated into many types of professions, including technical support and doctors.

Within this environment, paying an additional $40 per month for a mobile was expensive. Paying for incoming calls required a shift in mindset, and made people unwilling to give out their mobile number.

Mobile numbers are considered quite private, and many people will refuse to call a mobile if the landline number is available.

Compared with unlimited free email, ten cents for a text message is expensive, especially when many people cannot receive text messages. It took some time even after the carriers achieved SMS interoperability and mobile origination before a lot of people signed on. Messaging plans start at around $3 per month but can run up to $20.

Mobile Data Usage

US data use, like Japan's, should be separated into messaging and other use. The success of messaging-focused devices like the BlackBerry and the Danger's Sidekick suggest that there is a robust market for messaging services, but they should integrate into the well entrenched email and instant messaging ecosystem.

Americans, like Europeans and Japanese, like ring tones and other methods of customization. All also like games. The top mobile games in the US tend to match the top mobile games in the UK. The US market lags a little in penetration rates, likely due to:

- all the issues described in the earlier European section
- inexpensive Internet access on computers reducing the differential value of mobile access
- the computer-based advertising market, both email and Internet, being effective enough that mobile investment was not worthwhile
- service interoperability being harder than in Europe due to different standards.

Of course, these reasons feed each other: people were on the desktop-based Internet so content providers focused there; because content was on the desktop-based Internet people didn't move elsewhere.

2.3.4   Other Regions

Other parts of the world share characteristics with one of the three big markets listed above. Large parts of Latin America use GSM; China and India have adopted more of an American model of part CDMA and part GSM. Indian mobile phones are expected to have a Nokia-like user interface, whereas Chinese phones vary as much as Japanese phones do. The Korean market and users have as much sophistication as the Japanese market, with a similar industry structure.

If targeting one of these regions, you'll want to research your user base carefully. Different cultures have different behaviors and expectations of their mobile devices, and can interact with their technology very differently.

One dimension of variability is simply the degree to which users are likely to read the user guide. User research performed by many transnational product companies has indicated that many Americans never open the user manual, Italians are likely to toss it out with the packaging, and Germans are likely to read the entire thing before using the product. Similarly, Indian users are likely to read everything in the box, including the manual. A Chinese user, on the other hand, may lose face if caught reading the manual. Behavior varies across the world as well as from person to person.

These cultural differences are made clear by two differences of opinion I had with developers. One set of Korean developers thought I must surely be mistaken by insisting that names should be arranged by last name, and last name should be listed first by default for an American audience – they thought that certainly the 'first name' would be first. Another set of Indian developers argued that a particular button could readily control three modes of text input. After all, it was clear enough when you read about it in the manual. They were completely floored when I told them that just a small faction of their American users would even open the manual, and we adopted the simpler design.

In short, a bit of background research will give you a lot of information about historical factors affecting your target markets, and can suggest where user research will be most needed.

3

Mobile Devices

The current mobile device market has ill-defined and irrelevant market segments. There is an artificial distinction between 'phone', 'smart phone', and 'PDA'. This distinction appears to be based on the evolution of the device types rather than actual market segmentation.

Most industry analysts define 'PDA' as 'a handheld device with downloadable programs operated with a stylus but with no voice communications abilities', and a 'smart phone' as 'a mobile phone with advanced capabilities'. In standard industry practice, a PDA is a smart phone without voice capabilities. It's no wonder that PDA numbers are plummeting. Then again, a 'smart phone' is distinguished from a 'phone' by having advanced capabilities; this definition results in an unstable set of features.

The problem is exacerbated by Microsoft's branding of devices using their phone operating system as 'Smartphone'. I have seen reports that Microsoft coined the term, but the term actually long predates Microsoft's entry into the market. Many companies have defined 'feature phone' to mean a phone with data capabilities (as compared to a voice- and text-only phone) – the industry as a whole is likely to define feature phones as smart phones.

Clearly a better understanding of the mobile market is necessary.

3.1 A DEVICE TAXONOMY

Previously, we discussed characteristics of mobile users: interruptible, easily distracted, sociable, available, identifiable, and immersed in

Designing the Mobile User Experience   Barbara Ballard
© 2007 John Wiley & Sons, Ltd

their environment. Regardless of these commonalities, their needs and interests vary immensely.

These interests and needs affect users' choices in devices. An email-centric user might want a RIM BlackBerry or a Palm Treo, devices that have rich information services and interface but a less than ideal voice experience. An outside sales representative might live and die by a voice phone, and would prefer to relegate data services to a less than ideal experience. A medical doctor might need to see large amounts of information simultaneously, and could consider that large screen worth the cost of not being able to fit the device in a shirt pocket. A student immersed in social networking software would like a device focused on messaging and the web.

The mobile computing device market will not converge on a single physical form any more than the automobile market has converged on a single form. Devices will instead converge on a set of form factors based on market needs. The devices will fall into four classes:

- *general-purpose work*: multi-purpose devices, likely to be near the user while at work only
- *general-purpose entertainment*: multi-purpose devices with an entertainment focus, likely to be near the user when entertainment is acceptable
- *general-purpose communications and control*: multi-purpose, personal devices, used to communicate using voice and text as well as control things like home automation or finances
- *targeted*: devices intended for one or a very small number of tasks, with forms varying with their purpose.

Targeted devices are intended to do a very small number of tasks very well, and are available to the user in correspondingly more limited contexts. Such devices might be always present if they can become largely environmental: a wrist watch or an iPod can essentially be worn and forgotten; a clock is hung on the wall and does not require any sort of attention except when somebody needs to know the time.

User needs drive more than just feature sets, they also drive design decisions such as input method. A low-end phone works well with a scroll-and-select interface whereas a high-end phone might have a stylus interface. A device's primary purpose will affect its form; a game device, for example, is likely to be wider than tall and have several specialized game buttons. Different characteristics drive how the device is used and how best to design for a particular device.

3.1.1 General-Purpose Devices

General-purpose devices are intended to take full care of a specific market's needs. These devices are likely to be used frequently within their domain: work, entertainment, communications.

Since they have to support several functions, these devices will tend to coalesce into predictable computer-like forms: text input, cursor control, and a screen. The exact mix varies with device type. In contrast, targeted devices have fewer form restrictions and can instead be designed to perfectly match the tasks they support.

Work

Many, if not most, modern workers use some sort of computing device while working. While targeted devices include cash registers, inventory scanners, and ticket takers, the most common general-purpose computing device for work is likely the personal computer. However, the PC may not remain as ubiquitous as new mobile forms become available.

Computer manufacturers will continue to dominate the general-purpose work device market, with devices running operating systems similar to those on modern PCs. A more mobile device might have a tablet form, with a keyboard available but not required and multi-point touch or gestural input. It might have multiple screens, detachable from the device. It might readily connect with various environmental displays, ranging from projections and wall displays to private desk displays.

Because these are general-purpose work devices, they need to support screens large enough to view documents, forms, and the like. As a result, these devices will remain fairly large, with the size of the keyboard and screen limiting miniaturization. Even a foldable display will require space to use. For now, these devices are basically laptops or tablet computers with available operating systems.

While the decades-old promise of useful speech recognition has not yet been realized, its realization will not render keyboards obsolete. Speech recognition is useful for predictable text entry and commands. It will be best used in word processing situations and limited command set situations. It will not be particularly useful for changing labels for layers in Adobe PhotoShop, typing math functions, or precise character entry. It could potentially be useful for spreadsheet use, as long as there is a good method for error catching and correction. There may therefore be some sets of information workers who do not need a keyboard, but they will be in the minority.

Truly mobile workers have many of the same characteristics and challenges as the users described in this book. A delivery driver or meter reader might be interrupted by a person on the street. A sales person charged with tracking her company's inventory at stores will use her work device in a very sociable and readily distracted environment. General-purpose work devices, however, will tend to have operating systems based on full desktop operating systems; many of the device-imposed limitations will not apply.

Entertainment

General-purpose entertainment devices will have a cluster of entertainment features, based on market segment. One device might be media-based, with video and music prominently displayed. Another device might be game-based, with music and video as a secondary feature. A third device might be based on the written word, allowing the user to work pencil puzzles, read e-books, and browse the Internet.

While an entertainment device might be focused, it will still have add-on features. A multimedia device, focused on music and video, may have a book reader as an add-on feature. The written word device, focusing on ebooks, may have a music player and messaging, but likely not video.

The difference between a 'primary' and an 'add-on' feature is evidenced in the primary user interface of a device as well as the industrial design. Devices focused on voice communications have an obvious speaker, a numeric keypad, and a microphone; when numbers are typed at the standby screen, it assumes you are attempting a voice call. Devices focused on games will have game controllers as their physical inputs. On either of these devices, access to a web browser might be on a special button, but is more likely accessed through a menu system.

Add-on features are less easy to use due to the need to make the primary features easier. Any device that attempts to make all features equally easy to use will discover that the entire device is difficult to use.

Communications and Control: the PCD

When considering various communications technologies, it becomes clear that in industrialized societies, everybody has access to a communications device. Some people have a simple telephone, either landline or wireless. Others have communications access using voice over Internet protocol (VoIP) from a computer, specialized phone, or even mobile phone. Regardless of the form, communications devices are an increasingly important part of the lives of most people.

A communications device is a device whose *primary* purpose is communications. Certainly a personal computer is used to communicate, but communications can be considered a secondary purpose to general computing. Given that many people use their computers primarily for web surfing, email, and instant messaging, it becomes clear that some full-sized computers are communications devices.

The mobile communications device has a special role. It represents a person's always-available connection to the virtual world, both to information and to people. The importance of this connection was represented in the past by the prevalence of public telephones, which allowed connection to others while away from home.

The mobile communications device is so important, both to users and to mobile industry professionals, that we have given it a specific name: *personal communications device*, or PCD (see Figure 3.1).

A PCD is a mobile communications and control device. It is distinguished from other devices, particularly from full-sized computers, by being:

- *Personal*. The device generally belongs to one person, who will carry it either full-time or for a significant portion of time. This provides an 'always with you' experience that personal computers cannot match.
- *Communicative*. The device sends and receives messages. Currently, most PCDs use text messaging (Short Message Service, or SMS) and perhaps other messaging standards (such as Multimedia Messaging Service, or MMS) layered on top. This may not always be the case.
- *Handheld*. The device can readily be put in a pocket, worn on a waistband, or in rare cases strapped around the neck. Note that nestling a device such as a Tablet PC in the crook of one arm and then operating it with the other arm is not 'handheld', it is instead arm held and requires both hands.
- *Wakable*. The device can be awakened at a single touch by either the user or the network. A mobile phone will receive a text message even when it is 'asleep', or in standby state. Note that most computers, if they are asleep, cannot communicate with the network. This allows an 'always on' experience that personal computers cannot match.

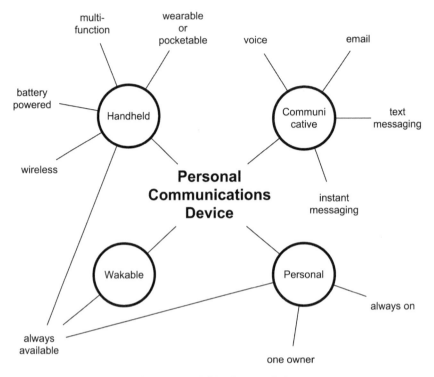

Figure 3.1   PCD characteristics

A communications and control device is almost always a mobile device. While desktop phones certainly could perform many of the activities mobile phones do, the user experience cost of doing so is too high compared to the value obtained. It is far easier to use desktop phones as targeted devices and move the remainder of the communications and control functions to the computer, which is likely sitting next to the phone and has better display and input capabilities.

3.1.2   Targeted Devices: the Information Appliance

Targeted devices are designed to help the user do a small number of tasks, and to do them well. Their form is thus highly variable and targeted at the exact device purpose. These devices include cameras, watches, televisions, radios, music players, credit card machines, automatic teller machines, and bar code scanners.

The functions targeted by these devices are frequently included in other devices. For example, most people have several clocks, and several more devices that include a clock as part of the function. Clocks are included in most computers, kitchen appliances, and car stereos, as well as being their own separate devices. Cameras are found in security systems, computers, and mobile phones, as well as in the hands of photographers and tourists worldwide. Users will tend to have multiple instances of the functions supported by a targeted device.

A targeted device is also known as an 'information appliance', a term coined by Jef Raskin[1] to mean a device designed to do a small set of information-focused tasks very well and be closely matched to the needs of the people using them. Raskin notes that these devices tend to be simple, always deal with information, and tend to share information.

Because the targeted device's simplicity of function, it cannot by itself provide the necessary ecosystem to support non-trivial data. A music device needs data to play. A camera is useless without a way to share or print pictures. An ATM is a sure route to bankruptcy without its connection to the bank's network. Thus all but the most trivial devices are part of an information ecosystem, and their data is shared with other devices and systems. Thus a typical characteristic of a targeted device is the need for reliable methods of data transfer. If it uses only proprietary data transfer methods it ties the user into a very small network, which could reduce the marketability or the usefulness of the device.

Given the likelihood that the user is already carrying a multipurpose device, there is little benefit to making a targeted device have lots of features. Any features that are not in the target set are going to be more difficult to use, or could possibly worsen the overall user experience. Features must be added to a targeted device with caution. Leave the job of a multi-function device to a device designed from the beginning to be a multi-function device. In other words, don't ask your watch to manage your investments.

One issue with targeted devices is the fact that developers frequently want to add on features. These add-on features can inhibit the overall user experience if not done carefully. For example, Apple added on a calendar view in its iPod. The existence of this feature simply uses the existing data connection with a computer and screen. This addition

---

[1] Donald Norman popularized the term in *The Invisible Computer: Why Good Products Can Fail, the Personal Computer Is So Complex and Information Appliances Are the Solution*, 1998, MIT Press (Cambridge, MA).

does not inhibit the use of the main function of the device: listening to music and other audio content.

Had Apple instead tried to add event entry into the calendar, at best event entry would not have been used much. The worst scenario would have been if Apple had decided to build a text input function to support add-on functions, which would have adversely affected the user experience for music.

Third-party developers are perhaps most notorious for demanding and building these add-on features, making the device into something it was not intended to be. These developers can have a disparately strong voice in product design, since device manufacturers understand that developers build device sales.

Targeted devices have fewer size issues than general-purpose mobile devices. The screen, if present, needs to be only as large as its data demands. Input mechanisms can be limited to only that which the target device demands, and need not be sized to support general-purpose text input. The shrinking size of music-only iPods, progressing to the size of a stick of gum, illustrates that screen size need not dominate the design.

Applications written for information appliances need to be written for the specific device or device family being targeted. This does not mean that some devices will not have general-purpose platforms such as Java ME or Linux, but instead that there may be significant customization of the platform. For example, MIDP 1 applications ran on Black-Berry devices, but could not use the device's navigation mechanism. To make a good MIDP application for BlackBerry, RIM's extensions must be used.

Historical

Abacuses and clocks are perhaps the earliest information appliances, storing changing information outside the brain. More recent examples include calculators, standalone word processors, cameras, and audio equipment. Most of these have evolved without the ability to share data with other devices, requiring paper or human to shift about data. They are therefore not stellar examples of information appliances, but are indeed targeted devices.

What we should learn from these devices is the enduring value some of them provided to society. Information tools changed navigation techniques, facilitated commerce, helped record history. All but one of the examples listed above remain in widespread use; even the mechanical abacus is still used in markets across parts of Asia. In contrast, the standalone word processor is not used much today, but a computer running a word processor looks so much like a standalone word processor that perhaps there is little need for a separate device.

Current

There is a broad array of targeted devices currently in the market. Targeted work devices, for example, are designed to support a set of similar job. Symbol Technologies designs devices surrounding inventory control, with extreme ruggedness and built-in scanners. Manufacturing processes are becoming more accessible to smaller organizations, with contract manufacturers willing to do an entire run of less than 10 000 units. This fact is leading to smaller and smaller companies being able to create truly custom devices. An early example of this phenomenon is the UPS Diad computer for UPS package delivery drivers, shown in Figures 3.2 and 3.3 and designed and built by Symbol Technologies.

This device has been so successful that UPS continually updates the design. The company has a separate device for warehouse package handlers, the ring scan. Note that the warehouse device apparently has less need for text entry, as there is a phone dial pad type of letter-to-number mapping on the warehouse device where the driver device has extra keys to support easier letter entry.

A more common information appliance is the iPod music player. Audio-only iPods do one thing well, and have a small number of extra features available. Video iPods, on the other hand, are more properly general-purpose entertainment devices.

Digital cameras are becoming pervasive as well. Like music players, cameras represent a function that could be, and often is, integrated into a multipurpose device. Nevertheless, the standalone devices still sell well. This is because the targeted devices provide a quality of experience and ease of use that cannot be matched by the necessary subsumption of feature access and use when it is included in a general-purpose device. A camera might need to be turned on, but once it is on pictures can be taken with a single key press. On a phone, the camera is accessible at best with a camera button, then the application is loaded, a picture can be taken, then menus are used to decide what to do with the picture.

Figure 3.2 UPS Diad IV targeted work device for drivers. Image downloaded from http://pressroom.ups.com

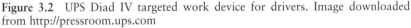

3.1.3 Ubiquitous Computing

Computing has expanded well beyond the terminal and mainframe model of the 1970s. The personal computer started the revolution of decentralizing data and some large portion of application functionality. Mobile devices extend this further, with connections both to personal computers and to servers. A complement to mobile computing is ubiquitous computing.

Ubiquitous computing is computing embedded in the user's environment. It is distinguished from 'computers' in that the devices are not personal computers, regardless of the hardware. Computing devices and displays recede into the environment, becoming invisible. Proponents sometimes call ubiquitous computing 'calm computing', as contrasted with desktop computers which actively demand the user's attention.

Ambient Devices makes devices geared towards information 'glance-ability', much like wall clocks. Their devices display information such as weather, stock prices, and so forth in an abstracted manner using physical devices. For example, the Ambient Umbrella pulses blue if rain is forecast for the day. Users of the 'dashboard', as seen in Figure 3.4, can subscribe to a large number of information feeds, including corporate data, and get information based on three analogue meters. The angle capitalizes on the eye's ability to quickly distinguish angles, particularly distinguishing vertical from other angles.

Various public information points can be considered to be early-stage ubiquitous computing, although the screen paradigm is still heavily embedded. These include ATMs, flight status displays, and kiosks. Note the similarity in scope of the targeted devices described above. The chief difference is that they are built from computers rather than from custom hardware.

Public information points are evolving to include services that directly interact with mobile devices via near-field communications or the Internet. Phone-pay vending machines and mobile-initiated printing have seen commercial deployment, certain applications can

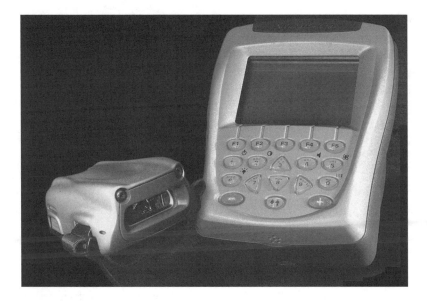

Figure 3.3 UPS Diad targeted work device for warehouse workers. Image downloaded from http://pressroom.ups.com

Figure 3.4 Ambient Devices' dashboard provides glanceable information in the environment without a computer. Image downloaded from http://www.ambientdevices.com/cat/gallery.html

be environmentally downloaded to mobile, and more general-purpose services are being developed in academic research laboratories.

Four major types of ubiquitous computing likely will be highly relevant to the mobile device ecosystem: pico nets, home servers, shared displays, and public interaction and download points.

Pico Nets

As users have more and more devices on their person, the need for sharing information amongst them becomes more important. The Bluetooth wireless technology was created to address this need, and infrared 'beaming' has been used in Palm and Windows CE devices for years.

The concept of a pico net, sometimes known as a personal area network, is the idea that all of a person's devices can share data with each other, automatically and wirelessly. Bluetooth was designed, for example, to support both wireless headsets and wider area network connection sharing. This vision has been slow to come to fruition largely due to carriers' hesitancy to open their networks to the resulting increase in use, which is a risk both to network integrity and profit.

However, pico nets of the future will share all sorts of data seamlessly, not just connectivity. One device that has only marginal use of a data store, such as an address book, might add access to that data store when it is available on the pico net. Thus a GPS device could quickly give directions to way points entered on the PCD, without major user input.

Home Servers

Home servers, such as Apple's Mac Mini and Microsoft's Media Center PC, will become more important. Home servers store videos, music, pictures, and data backups, serve content to various parts of the house, coordinate data between different users, and run home automation systems such as security cameras. Mobile devices can store subsets of this content, and can also manage the servers – and hence the home – remotely.

Future applications include answering the door from the mobile phone, regardless of whether the user is at home. A delivery driver knocks on the door, triggering an MMS with a picture of the driver and perhaps a second picture of the driveway or street sent to the homeowner's phone. The homeowner can then initiate a voice over IP connection to the front door and tell the driver to leave the package.

Shared Displays

A solution to the too-small screen problem is to simply connect the device to a larger screen. To some extent this is done in conference rooms with projector displays, but a variety of implementations are possible. A conference room table or wall could display content directly from the mobile device. Add a bit of interactivity and group access, and a sophisticated collaborative application could result.

Phone booths of the future could provide a degree of visual privacy for a display, allowing users to interact either via voice or keyboard. Similarly, walls in private homes could display aesthetically pleasing content until somebody wanted to interact with their device with a large screen.

Public Interaction Points

An extension of public displays is more fully featured interaction points. These could allow the user's device to pull down data and also push information back.

One student project suggested using near-field communications to intelligently display airport location information: as a user approached a gate information display, a 'you are here' indicator unique to the user is displayed. As the user gets closer, the icon grows larger; as the user moves right or left the icon follows. The icon contains gate information and number of minutes until boarding, or some similar information. This information could simultaneously be pushed to the mobile device.

3.2 ANATOMY OF THE PCD

Of all the general-purpose devices, the personal communications device is the most ubiquitous. It is always carried by its owner, which has several important implications in its design. To make an application available to as many people as possible, it will need to be delivered on a personal communications device.

The PCD is personal, communicative, handheld, and wakable. As a personal device, it is not likely to be shared with others. As a handheld device, it is small, battery powered, and wireless. As a communications device, it is usually on and connected. It is turned off only in rare situations, and connectivity disappears only temporarily.

The PCD is also a general purpose device. It therefore has the four main components of any general-purpose device: display mechanism, focus control, text input, and development platform. It has several other characteristics as well.

3.2.1 The Carry Principle

While users will frequently have their general-purpose work device, they will not when not at work. In contrast, a PCD is always with the user. This fact has profound implications on device and service design, and will be explored further in the Principles chapter.

The fact that a PCD is carried with the user all the time means it is multi-functional. Users will allow for a certain difficulty of use for the privilege of having the device readily portable. This is akin to a

Swiss Army knife: its blades and tools are serviceable, but they are not appropriate for heavy use. People carry a Swiss Army knife sooner than a set of knives for its convenience and portability. Similarly, the PCD provides an array of voice communications, text communications, house control, applications, etc. It is the most personalized device.

The Carry Principle dictates the characteristics in a successful PCD.

3.2.2 Input Mechanisms

Input mechanisms include a variety of methods for getting data onto a device. Mechanisms can be categorized into focus (cursor) control, commands, text or character entry, environmental data entry, and other-computer data entry or access.

Focus Control

Perhaps the category with the most fundamental impact on application design is focus control. This is the method that the device uses to decide the object to which to direct any user input, and the most common methods are stylus and scroll-and-select. A stylus is similar to a mouse, but has no cursor and does not have the ability to access multiple commands without complex actions like press-and-hold or the very difficult double-tap.

A scroll-and-select mechanism has up and down and usually left and right controls and a select button. While many devices use a 'five-way rocker' with each of the above controls, others use a 'jog dial' or other physical mechanisms. Scroll-and-select works rather like tab and shift-tab on a computer, with some acceleration of navigation available in certain situations. Some phones also support a scroll control for screen-by-screen movement, usually via the volume keys.

Other focus controls are possible. Accelerometers can navigate through a series of pictures with a wrist gesture or perform other actions. Speech can select an object on the screen, although this is fraught with user experience and technical problems. Focus can also be controlled using keyboard shortcuts, such as numbered list items.

Commands

A more subtle mechanism category is commands, the various methods beyond select and activate the device uses to perform actions. Hardware buttons to activate programs, such as a camera button, are the key example. Some phones also have a Back or CLR key, which is used for both aiding character entry and navigation between pages.

Softkeys are on-screen buttons that can be quickly accessed by unlabeled hardware buttons. They provide context-sensitive commands in a hybrid of software and hardware button. Devices vary in how they implement softkeys, and different platforms have varying access to softkeys. Devices may use, among many options:

- *Nokia-style Options/Back softkeys.* Any contextual controls are in a menu launched by the Options button. Back becomes Cancel in certain contexts. These phones do not have separate back buttons.
- *Simple softkeys*, with two or three virtual buttons and the corresponding number of hardware buttons. The virtual buttons have labels indicating what actions the hardware button will initiate. Some phones have separate select buttons, others do not. Either type of phone may have parts of the user interface in which a softkey is used as a select button.
- *Samsung-style OK/Menu softkeys.* Samsung has used its OK and Menu hardware buttons to access softkeys. The OK button is also the device's select button, so this is essentially a one-softkey design.
- *Scrolling softkeys* do not have physical softkey buttons, but instead have left and right scrolling through a list of actions available for the currently selected screen or object. The select button always operates on the action list, never on the object directly.

Third-party software that is burned into the device's memory may not follow the conventions found in the remainder of the device. Browsers in particular are likely to break with the conventions, particularly in their use of softkeys, because the standards have evolved to drop softkey support.

Speech commands have been present in mobile phones for years, but they are infrequently used by end users. People consider the feature when making a purchase but find themselves rarely if ever using it. As processor capability and amount of content increases, speech recognition will become an increasingly important mechanism for navigating and acting on content.

Mobile search, both of device content and Internet content, is likely to be best achieved via speech input, with a combination of natural language search and robust search results based on all likely uttered words, not just the most likely. The natural language search increases accuracy by giving the recognition engine a predictable grammar, and the use of multiple possible words for an utterance significantly reduces the negative impact of misrecognitions.

Text and Character Entry

The input mechanism that has garnered the most attention is text and character entry. Mobile phones are notorious for having difficult text entry, although some users gain significant speed. The type of text entry is partially dependent on the intended use of the device. For most devices, voice calls need to be dialed using one hand only. This has limited voice-targeted devices to a standard 12-button keypad, and its variations. Devices more targeted at messaging can support a two-handed text entry mechanism.

A one-handed text entry mechanism will not be a keyboard-based device simply because if the keyboard is shrunk far enough to have all its keys reachable by a hand holding the device, then the keys become too small to be operable by a thumb.

The normal one-handed input mechanism is some variant of the standard 12-button keypad, including * and #. Normally, triple tap is used to access letters on each key: a 'r' requires three presses of the 7 button. A two-tap mechanism is also possible: a 'r' requires a press of the 7 button, then a press of the 3 button for which letter it is on the key. This mechanism is slightly faster, but is not widely adopted.

Recent years have seen a variety of one-handed keyboard alternatives become available. The Fastap keyboard has letters nestled in between the letters. Accidental activations are avoided by not having the numbers be buttons at all; instead numbers are activated by chording the surrounding letter buttons. This chording is invisible to the user and does not require precision from the user.

Other one-handed text input mechanisms have come on the market. Some are doomed because they don't solve the fat-finger problem. Others use some version of simultaneous button press (chording) to activate single characters. Gestures of various sorts, such as using a force stick to 'write' letters, are also available. These mechanisms are likely to stay in niche markets, such as PCDs with very little emphasis on text input.

Two-handed text input solutions fall into the categories of thumb keyboards, handwriting recognition, and virtual keyboards. Thumb keyboards are found on the BlackBerry and Palm Treo devices, amongst others, and have buttons operable by people with medium or small sized fingers. Fingernails can also get in the way.

Handwriting recognition is provided by a number of companies with varying success. Some users can get very high recognition rates; other users have a harder time.

Virtual keyboards operated by stylus vary widely. Some are merely QWERTY layouts, whereas others build in letter and word prediction with a variable display, but WordLogic uses an intelligent combination of the two. Users start typing with a standard QWERTY keyboard, have the most likely next letters highlighted, and have complete words displayed to the left. Further, a simple gesture function allows users to build parts of words. A long word frequently is written with two to five taps. These may be built in to a device, or may have been downloaded as an additional tool.

Some solutions require not only two hands, but a surface. These include any full-sized keyboard, whether rigid, rolled, or virtual, as both hands are used for input and not able to hold the device while doing so.

Complementing the hardware for many devices are letter or word prediction programs. A character prediction method is very useful on a 12-button keypad, as it reduces keystrokes by more than half; a version of the same program can be used to increase accuracy in handwriting recognition. A word completion program, which is separate from character prediction, suggests words that match the currently entered first characters. Such programs are useful for even the easiest of mobile text entry mechanisms.

Some platforms, particularly browsers, do not have access to the device's prediction programs. Other platforms have only rudimentary access: the user turns prediction on or off for the entire platform at once. For devices in which the application platform has careful management of prediction programs is necessary, as some fields do not lend themselves to dictionaries whereas others do. When using the platform, you may not have access to the prediction programs.

Environmental Data

Access to information beyond the confines of the device is one of the places where mobile devices are actually more capable than their desktop counterparts. Cameras, RFID readers, various location technologies, thermometers, and any number of other potential input mechanisms gather information from the environment and help understand user context.

The camera is the most prevalent such input mechanism. Its use goes beyond taking and sharing photos with friends, and progresses to bar code recognition and generic image recognition of products, movie posters, and people. Taking a picture of a menu could add ideas to your recipe box. A picture of a meal could help you record calorie, fat, sugar, and carbohydrate consumption.

Expect the camera to be very important in mobile search, with comparison shopping becoming useful as products similar to the item pictured are found. Previous versions of comparison shopping looked only at items with the same model number, and major retailers secured models with different numbers but the same characteristics. Current versions, accessible by voice, SMS, on-device application, and web, can suggest similar products.

Other Computers

Other computers also provide critical data. Servers are obvious, but ubiquitous computing systems and other devices of the user's personal network also provide useful information. A future version of the iPod, for example, might be connected via Bluetooth to a phone. When the phone rings, the iPod would pause the music, switch to phone headset mode, and allow the user to answer the call without changing earpieces. Such a feature would replicate similar features in an integrated device.

Synchronization, either with the user's own computers or with a commercial server, also provides input. There is a growing trend towards accessing media content, including both music and television, from the user's home content library rather than accessing content directly. This type of input is sometimes known as 'place-shifting' when live television from the home is viewed on a mobile device, and 'time-shifting' if home-stored content is viewed at different times.

3.2.3   Output Mechanisms

Screens are the most obvious of output mechanisms, with the LCD as the most common and other technologies in various stages of productization. While these are the most obvious, the technology actually impacts design of applications.

The LCD screen will become less and less popular due to power and cost issues. These screens are rigid, have significant polarization issues, and require significant backlighting to become visible in sunlight. With an LCD screen intended to be used outside, all information-laden graphics need to be high-contrast, with thick lines. Text color must be high contrast with its background. On the other hand, the polarization means that the screen is more difficult to see from the side, making information more secure from casual observers.

The LCD backlighting introduces challenges for the user for applications with low interactivity. The screen will fade after a few seconds of lack of user inputs. While this is generally a setting that can be changed by the user, it falls into the category of things rarely found by the majority of the users. This leaves the user introducing spurious inputs to keep the screen lit while reading or studying the screen.

OLEDs[2] are made with a radically different manufacturing process. The OLED pixels emit light directly, giving them better visibility in sunlight, reduced power consumption, and no polarization issues. OLEDs have not taken over from LCDs because they have a shorter life; researchers are addressing the issue. These screens give the designer a broader range of color choices and allow for more subtlety in design.

Electronic paper[3] displays have a set of balls as pixels. Each ball contains a smaller ball[4] with two colors; the electric charge tells the inside ball which color side to display. These displays require low power to change, and no power to maintain the display. They can only change approximately four times a second, making them inappropriate for highly interactive displays. They have almost as good readability as newspaper. If designing for this type of display, eliminate animations and reduce screen changes. A clock on the outside of a phone, for example, should update once a minute.

Electrowetting[5] displays use an electric field to decide whether a colored oil covers or doesn't cover the substrate. These displays have excellent color and low power consumption. Most of the manufacturing process is the same as LCDs, which should allow it to quickly enjoy economies of scale and have similar costs, but the technology remains very new. Unlike electronic paper displays, they can also be changed at video speeds.

---

[2] Universal Displays is a major manufacturer.
[3] E-Ink is the primary technology owner.
[4] An alternative version of the technology has several balls inside a colored liquid in the larger ball. The colored liquid provides the color, obscured by the balls at need.
[5] Liquavista is the technology owner.

OLEDs, electrowetting, and electronic paper can be made on flexible displays and require less backlighting, making them use less power.

For devices with multiple displays, we expect status displays to be electronic paper, and video displays to be OLED or electrowetting. Thus many issues associated with graphic design for mobile phones will be abated. Glare issues are reduced with electronic paper. Electrowetting and OLED allows for beautiful color with broad angles of view. All these technologies enjoy lower power consumption, which will allow for longer use between battery charges.

Various connection technologies such as Bluetooth, Wi-Fi, WiMAX, and infrared can be considered output methods, but are instead described under connection characteristics. These connections can send data to other computers, including the environment, servers, nearby devices, or other devices within the pico net.

Various speakers can also display data. These include the built-in phone speaker, a speaker phone, and an earpiece. The vibrator, if present, is also an output mechanism and is accessible by some application technologies.

3.2.4   Technologies

PCDs support a variety of application technologies, each with different strengths and weaknesses.

Browsers

Most devices have a browser of some type, provided by Openwave, Nokia, Access, AU Systems, Opera, or some other provider. This browser, if found outside of Japan, is likely to support XHTML Basic or XHTML MP as its primary markup language; a Japanese browser may support cHTML (compact HTML) instead. All are restricted versions of standard HTML/XHTML. XHTML browsers will support CSS whereas cHTML browsers will need styling defined inline.

Some newer browsers also support scripting and even AJAX (Asynchronous Javascript And XML). In general, any prefetching that a web application can do will improve the application responsiveness and hence the overall user experience, so these technologies will become important as they spread.

Messaging

Devices also have a variety of messaging capabilities. SMS text messaging is nearly ubiquitous, although it took a few years for US providers to make it interoperable and two-way. MMS (Multimedia Messaging Service) allows for the transfer of pictures, text, and sound. It is hampered by cost and interoperability issues. Mobile blogging applications may reduce the attractiveness of MMS, but that remains to be seen.

Voice SMS allows the user to record a voice message and send it to another mobile user. It is essentially a voicemail message that does not attempt to reach the user directly first, with the capability to send messages to groups.

Application Platforms

A device is also likely to have one or more application platforms that allow development of a local application. These can be divided into native or targeted platforms, broad availability platforms, and limited availability platforms.

Java ME is perhaps the most widely deployed of the broad availability platforms. Its creator, Sun, worked towards a 'write once, run anywhere' solution and designed Java ME to be able to run the same program on devices with different capabilities. As so frequently happens, the reality did not meet the promise due to poor implementation of the application environment user interface and varying technology implementation.

Flash Lite will have broader and broader availability, although it is currently limited. It combines scalable vector graphics (SVG) with ActionScript, which is based on ECMAScript. Flash Lite allows rapid development of applications for specific devices, but does not provide any method of automatically changing application appearance based on device capability at the device: all optimization must be done by the developer at design time. Adobe's promise of fast application development across all devices should be tempered by the reality of device variances, but the problems will not be as profound as they are with Java ME.

BREW was designed for broad deployment, but is on the Qualcomm CDMA chipset so is not natively available on GSM devices. There do exist a few GSM deployments. It is, to the user, similar to Java

ME except the applications run faster. Deploying a BREW application requires the carrier to sell it.

uiOne (formerly Trigenix) was designed to allow users to customize their device's native user interface. Theoretically this customization could remove feature access from the device; in practice only the first level or two of the phone is customized with graphics, animation, ordering, and sounds. Clearly the operator allowing a uiOne application will want to ensure that all the operator's money-making ventures (messaging, voice, browser, etc.) and support needs (settings) are as usable as they were before the customization.

uiOne now is part of BREW. As BREW is actually a productization of Qualcomm's internal development platform for device user interface, this opens up the possibility of some very sophisticated services. Expect access to such services solely through the operators, so only organizations with strong carrier ties will be able to take advantage.

Python and OPL (Open Programming Language, formerly Organizer Programming Language) are languages for developing for the Symbian platform. They are each interpreted, making them slower than compiled languages such as BREW and C++.

Linux applications can run on an array of devices, but may require significant recoding for different versions of Linux.

Purely native application environments include Symbian C++, PalmOS, Linux, and MS eMbedded Visual C++. These are compiled applications. They have deep access to a device's capabilities, but limited cross-device applicability.

Media Players

Media is becoming ever more important. Video distribution has traditionally occurred on mobile devices point-to-point, with a unique connection between the operator and the individual user. While this allows for highly customized experiences, it is not bandwidth friendly and is limited in its scalability. Broadcast solutions will be available soon, and devices may be able to record segments for local playback and forwarding.

Person-to-person forwarding of video clips, like pictures, will become more prevalent. Some mobile marketing firms are in fact counting on this, and profess expertise in viral marketing. They believe that they can create advertising content, perhaps embedded in something popular, that will be forwarded from person to person without sustained investment from the advertising firm.

The central standards for media is 3GPP (Third Generation Partnership Project) and 3GPP2 (for CDMA), which define a mobile platform for MPEG-4 formats. Specific formats include AAC and AMR audio (plus QCELP for 3GPP2) and MPEG-4 and H.263 video. Content production tools such as QuickTime Pro readily provide the correct formats.

3.2.5 Connection Characteristics

The PCD is a wireless device. As such, it has:

- power consumption concerns
- inconsistent coverage
- speeds slower than prevalent land line speeds
- limited coverage area and hence potential roaming charges
- latency in connection, particularly for establishing the connection.

These characteristics impact application design. For example, an application whose data must be present on the device, such as a calendar or contacts, should not be a pure browser application. For that and similar needs, a local application with network access is preferred.

3.2.6 Standby Screen

The standby screen is the main device screen, before the user has interacted with it. It provides valuable real estate for branding, advertising, and personalization.

User interfaces can be defined to have applications or actions available on the main screen, or applications only available with the press of a hardware button. Most devices assume that if the user starts typing numbers, that a voice call is to be made; this leaves the main screen free.

Currently, wallpapers for screen customization are popular and lucrative. In the future, users may be allowed to have reduced cost of using the phone in exchange for branding on their main screen.

Application platforms such as uiOne allow for significant customization of the standby screen. Sprint users, for example, can download themes that have four links on the main page in addition to softkey links to Contacts and Favorites.

4

Selecting Application Technologies

Most business goals can be accomplished by building an application using a variety of technology combinations. A search application can be accessed using voice, SMS, MMS (camera and visual search), web, or an on-device local application. Each of these has different user experiences, device portability, user coverage, and in general overall user experience.

While most designers do not get an opportunity to select the technologies that will be used, marketers do, and this is the first decision that affects the user experience. The ideal scenario, of course, is to get the content to as many people as possible, with as good a user experience as possible, with as little development as possible. This chapter provides a framework for making these decisions.

Many of the identified technologies could quite reasonably be on devices beyond personal communications devices. A digital camera, for example, could have network connectivity via Bluetooth, Wi-Fi, or cellular, and a development environment like Java ME MIDP. This combination would allow direct access to various photo-sharing applications, such as a newspaper photojournalist image submission site, a photo sharing site like Flickr, or a blogging application. This array of applications would not be supportable by the camera manufacturer directly in the software, but an API could make the camera more attractive to customers once software is available.

Selecting a platform, or combination of platforms, clearly needs to be done with the full collaboration of your technical staff, who will be

Designing the Mobile User Experience   Barbara Ballard
© 2007 John Wiley & Sons, Ltd

considering a number of factors not mentioned here. This may include application size limitations (which will vary over time, and thus are not described here), in-house expertise, technical capabilities, and so forth. The business members of your team will also consider what platforms are supported by the desired partner device manufacturers or carriers.

Platforms, and their capabilities, will continue to change. These changes are usually improvements, but occasionally capabilities are removed in the name of compatibility. This constant state of change implies that by the time this book is printed, specific platform data is likely to have changed; the analysis points raised in this chapter will remain the same.

4.1 INPUT MODALITIES

The method of input – the phone keypad and focus control – may seem obvious. There are, however, other options. Most PCDs are optimized for voice, and voice-over-IP (VoIP) allows the device to connect to a server using voice, without establishing a separate voice call. This fact will end up impacting mobile applications profoundly by allowing voice and sound as data to be transferred over the same connection as all other data, eliminating the need for a separate voice call. Cameras are also excellent input devices, allowing for a number of visual-based applications. Few, if any, applications will be operated purely by camera.

With the continuing evolution of device capabilities in mind, we observe that a device can use as input visual, auditory (speech or other sound), or touch from the environment, much like humans. While other inputs, such as location and temperature are possible, for most applications they will not be the primary method of controlling the application. We thus relegate these to 'Supplemental Technologies', below.

4.1.1 Buttons

Most applications discussed in this book are operated by pressing the physical buttons on the device, or operating a stylus to press virtual buttons and capture handwriting. This type of input is extremely familiar from personal computer applications. While most applications use buttons for input and control, the phone is designed to be first and foremost a voice device, with buttons only to perform necessary supplemental tasks.

4.1.2 Speech

Speech can be a natural input mechanism for a number of types of applications. Natural does not, however, mean easy to implement, design, or use. Decades of research and failed products have proven this.

Challenges associated with speech input include input inaccuracies due to difficult accents, mismatches between spoken speech and sampling frequencies, and grammar design difficulties. Speech systems can also leave a user feeling powerless if her utterances are rarely recognized, which can readily happen in mobile phone use environments due to environmental noise.

Speech also introduces further privacy and politeness issues. Mobile phone conversations with humans tend to involve a more projected voice than do in-person conversations with a human in the same environment; this fact has created ongoing resentment against others using mobile phones in public places. When a user talks with a machine, they are likely to project their voice even further, suggesting that the entire train station or coffee shop will know exactly which application is being used.

4.1.3 Speech + Buttons

We have speech combined with buttons as a separate input modality due to both the difference in user behavior if using both speech and buttons, and the differences in application platforms possible. Examples of applications using both speech and buttons include voice response systems like voicemail that ask the user to press a button to perform an action, and multimodal applications that allow voice to supplement a mostly visual application.

Voice-over-IP (VoIP) is going to increase the possibility of designing applications with both speech and buttons. Earlier technologies required separate connections for voice and data, and one session had to be ended before another began. As VoIP spreads, speech and data can be accomplished simultaneously, creating the capability for fully integrated multimodal applications. Companies such as V-Enable have been working towards this vision, creating a server environment that can recognize speech commands given to an otherwise button-driven interface.

Using speech as an alternative interface allows users to decide which input mechanism is right in their current context. If recognition fails, they can revert to buttons for input.

4.1.4   Visual + Buttons

The camera is an input device whose importance will increase over time. It is one of the key methods of inputting environmental data into the phone. A well-designed application could use the camera to input:

- a bar code or other visual identifying symbol displayed anywhere in the environment, allowing for a quick retrieval of specific product information or marketing interaction
- an advertisement as a whole, such as a movie poster, for quick retrieval of product information or marketing interaction
- a face, both for addition to the user's phone book but also as a source for tagging for photo classification
- a car or even its VIN, to get history of the vehicle before purchasing it
- a face, to see whether it is in a list such as the sexual predator list, missing children registry, or simply within one's social network.

There are numerous uses of a camera. Cameras will not be put in widespread use as input devices until the devices support the taking of a picture and then choosing what to do with it, as the amount of time it takes an application to load is likely going to inhibit any rapid use.[1]

4.2   INTERACTION RESPONSIVENESS

The responsiveness demanded by your users and application should match what the platform can provide. The fastest response is found in well-designed compiled applications with memory management being run directly on the processor rather than in a virtual environment.

---

[1] A simple extension to a platform such as Java ME could be a 'picture action', which is a label and command to be added to the device's menu of things to do with a picture. This menu already includes the ability to send the picture to contacts, but could easily include a command like 'Find Product' or 'Comparison Shop' or 'Missing Child?', each attached to different applications. Selecting the command would launch the application on the device.

The slowest response is found in messaging applications, which can best be described as asynchronous. As always, the quality of code implementation has much to do with responsiveness.

- Asynchronous applications do sometimes have inherent advantages over their more interactive counterparts. Their asynchronous nature allows for user and device interruptibility, as the application itself is interruptible. Further, the results are stored and displayed locally, in a predictable location, and the application need not be 'running' to get the results. This makes it particularly good for temporary content that will be accessed a few times, such as static directions.
- Fast applications are run locally, are directly compiled, and run directly on the device's native instruction set.
- Medium-speed applications are run locally, but may be interpreted or may run in a virtual player or environment. Application loading is likely to be slow, and interaction will not be as responsive as might be desirable in many action game applications. Online resources about the platform will extensively discuss methods for speeding up applications.
- Slow-speed applications generally have significant network delays as the application waits for information. While any application with network access may experience network delays unless the need for the information is accurately forecasted, browser applications without AJAX technologies will experience these delays with every interaction.

4.3 DATA STORAGE LOCATIONS

Many applications store user data beyond temporary interactions, whereas others do not. The user's need for persistent storage, either locally or on a server, varies based on the nature of the data and the application's requirements. A calendar needs to be available even when the network is gone.

The location of data storage is more important for users in areas with inconsistent coverage, such as US users,[2] but it is relevant for

---

[2] Data coverage outside of metropolitan and other highly traveled areas is quite spotty, with miles of Interstate highways in the West with voice coverage only. While the majority of the population lives in an area with good coverage, much of the land mass and many travel destinations do not have good coverage. Additionally even in metropolitan areas there are coverage holes.

many other users. If the application needs to be available while out of coverage, such as on an airplane or in a tunnel, it should be stored locally. For these reasons, there will continue to be a need for both browser-based and locally based applications.

Messaging applications are transient in nature, although the user can choose to store the results locally. These are good for situations in which the user's past input, behavior, or results need not be considered. However, a messaging connection to a server-based application, such as messaging access to a PayPal account, can be used to overcome the transient nature and instead have remote data storage.

Local data storage can have its own issues. Since PCDs can be readily lost or swapped, any local data needs to have a backup stored on the user's desktop computer or a server. This process needs to be carefully managed.

Remote data storage has high reliability and addresses the issues associated with loss of device. It has challenges associated with network access, as discussed above.

4.4 DISPLAY MODALITY

Devices can display information using aural and visual displays. Some platforms and devices also allow access to the vibration function.

Aural displays can be played via the ear piece or the speaker. In many situations, sound played via the speaker during an application will be disabled due to privacy or politeness issues. Sound played via the earpiece of course makes the user's ability to see and hear the display simultaneously more challenging.

Applications thus tend to segregate themselves into those with sounds playing key roles, and those with sounds playing supplemental roles. Expect use of the former to be somewhat limited by user situation.

Visual displays are common, and expected for all but speech applications. *Tactile* displays, such as the vibrator, are accessible by some platforms. This is notorious for using battery life quickly, but is important for getting the user's attention in noisy environments.

4.5 SUPPLEMENTAL TECHNOLOGIES

Some platforms provide access to a device's capabilities beyond button presses, display, and perhaps the speaker. Of course, device support for such technologies is highly variable, but these technologies can greatly enhance the user experience.

Location, as measured by global positioning system (GPS) or other technologies, is fraught with privacy issues, enough that many users will not have it turned on. Java ME, BREW, and native platforms have varying degrees of access to the information, depending on what portions of the platform's capabilities the device has enabled. Java in particular has varying implementation of JSR[3] 179, the location interface module.

Carriers (or devices) may allow access to location, but only for registered applications. They may allow access, but not at a precise level. They may even charge for access to the user's location. Creation of location-based services (LBS) requires an in-depth analysis of market, carrier, and device capabilities; you will likely find that only one platform is even an option.

Wireless connections to other devices are also sporadically supported. Java and BREW applications can access Bluetooth local networking if the device supports it. Some native platforms also allow access. Palm and PocketPC allow prolific access to the infrared port but cross-device platforms have no access.

User data on the device, such as the calendar and contacts, is accessible by native platforms, BREW, and the Java ME PDA profile specified in JSR 75. This data can reduce the need for text entry, provide a local display for online data, and in several other ways enhance the user experience for certain applications. For a messaging application, such access is very important.

Some platforms allow the storage of small bits of data by applications; cookies are the prime example, bookmarks could be considered a specialized version. Flash and Flash Lite both allow such storage. More capable platforms have this capability with the standard file system.

Devices will have an increasing number of methods of display. Vibration is often available on current devices; many 'clamshell' phones have a secondary display that can be used by some platforms. Future devices may have projected displays or even odor-generating displays. As always, native applications have the greatest access to such features. BREW has vibrator and secondary screen access; Java ME, in MIDP 1 and 2, has only vibrator access and then only for some devices.

---

[3] Java Specification Request, the specification for different aspects of Java. The Mobile Information Device Profile, for example, is JSR 37. MIDP 2.0 is JSR 118. Theoretically, the JSRs that a device supports lets the developer know what is and is not possible on that device; in practice different devices implement a given JSR differently.

Devices will also have an increasing number of data sources locally available. A thermometer or glucose meter might provide information about health issues. Device manufacturers are experimenting with accelerometers to allow users to gesture to control the devices. Watch platform capabilities to see what access is available; expect it in native platforms first.

4.6 DISTRIBUTION METHODS

Distribution methods include both broadcast and point-to-point models. The former will distribute content at low costs, the latter allows for on-demand media but suffers from scalability problems. The future is likely to see a mix of the two, based on media type and user behavior. See Chapter 8 for more information about distributing applications.

4.6.1 Cost of Deployment

A platform's cost to deploy applications is a function of programming complexity, rendering differences among targeted devices, the carrier, and cost of getting the application into sales channels.

Programming complexity is inversely related to the platform's access to device capabilities. In general, the same things that make a platform powerful will make it more complex to code. Further, greater access to device capabilities by a specific application seriously increases the impact of varying device capabilities and rendering algorithms.

The same technology displayed on different devices will frequently render very differently. This problem will continue to exist due to varying user needs across market segments. The problem is compounded further when an application needs to render in different (but equivalent) technologies, such as Palm and Windows Mobile.

For some platforms, rendering engines are available. These engines optimize generic mobile content for display on devices with different capabilities. A voice SMS engine, for example, would send voice SMS to devices that support it, and a SMS with a callback number for devices that do not support voice SMS.

Rendering engines are limited in capability, and are best used for platforms with limited interactivity on the device or limited rendering differences. All such engines successfully capture display size; most also capture device capabilities. Few capture device user interface differences or differences in how certain commands are interpreted. Avoid engines that purport to render to mobile or desktop environments with the same base code; see 'Class-based Design' in Chapter 5 for more details. Definitely use caution when selecting a rendering engine that purports to create applications on different platforms based on the same code: always expect human intervention in translating applications between platforms.

Good rendering engines are available for web sites, SMS, and MMS. Flash Lite allows rapid recompiling of designs for different device capabilities, although it is limited in what capabilities it can access. Java ME rendering differences can be partially addressed using WURFL and other technologies; again see 'Class-based Design' in Chapter 5 for details.

4.6.2 Sales Channels

Different platforms have different advantages and challenges with regard to sales channels. These differences are largely due to the carriers' business models and users' willingness to pay for services.

In the United States, for example, SMS has seen slow adoption, particularly among adults.[4] The country relies far more on email and instant messaging and suffered from carriers creating barriers to interoperability. Thus reliance on SMS for delivery, except for various youth markets, will limit penetration compared to voice. However, SMS is perhaps the most commonly used platform as it has the greatest coverage and its use is growing.

Web browsers are very commonly available on devices, but the user has to be able to both find the browser on the phone (difficult on devices including a Motorola RAZR from Verizon) and have a data plan that supports browser use in a reasonable fashion. Cingular's data plans as of March 2006 had 1 megabyte transfer per month charged

---

[4] US adoption has lagged behind European adoption for a variety of reasons. First, European operators originally did not expect SMS to be popular, so they priced it for rapid adoption. Second, high telecommunications costs in Europe meant that computer and Internet penetration, particularly at home, lagged the US. These two facts made SMS a spectacular deal. US carriers, seeing European SMS success, priced SMS at more of a premium, while email and instant messaging penetration was quite high among teens and the population in general. Couple this with different pricing models in the US, such as the recipient must pay to receive a message, and cross-carrier incompatibility, and the recipe for slow US adoption becomes obvious.

at a US$5 monthly fee; larger amounts of data cost $15; unlimited browser use was $20. Delivering a service using browsing technologies would be limited only to users who were willing to pay an additional $20 per month, all going to the carrier. Any content charges would raise the user bill even further.

The 'walled garden' refers to a carrier's prohibition of content beyond what the carrier has authorized and contracted for. This practice was predominant in 1999, and still exists with many carriers in 2006. The original intent, at least at Sprint, was to protect business relationships and maintain a minimally usable user experience. As the mobile web has grown and more content has become available, the original intent is no longer valid.

Verizon and Cingular both maintain their walled gardens in 2006. Thus Verizon does not allow URLs in text messages to be clickable: the user would have to manually type the URL into the browser. Cingular has a clause in its user agreements stating that the user will not visit sites outside Cingular's properties. The access that Sprint Nextel gives to their customers to sites outside the 'garden' varies, but the user can always type an arbitrary URL; if it is compatible with the mobile it will work. Regardless, many users cannot figure out how to enter a URL, so the on-deck content is most accessed.

Thus a web service would be available to Sprint and most European customers without special relationships, but not to Verizon and Cingular until they either open their networks or your organization has a business relationship with them, putting you on their portal. Check carrier policies in your market for a good understanding of the challenges you will face.

Even assuming that the networks are open, positioning on a carrier's portal may be extremely useful for promoting your service. Certainly the history of desktop portals suggests this to be the case, with deals associated with the placement of content. Entering a URL on a phone is more challenging than entering a URL with a full-sized keyboard, so we should expect this trend to continue. Note that only web services can be placed on the portal, as carriers are unlikely to place a link to a downloadable application as a main link on a space-constrained portal.

Downloaded applications are acquired, by users, from three main locations: the carrier's store, a third-party store such as Handango, or the software provider's own site. For the most part, third-party stores appear to carry more native applications than cross-device applications written in Java or BREW. Indeed, BREW's business model requires carrier involvement for the sales process.

A SMS product does not necessarily involve the carrier, and can be monetized directly using premium SMS and short codes. It is not without its limitations. PayPal's re-entrance into the mobile payment arena is likely to inhibit the use of premium SMS in the United States due to a greater familiarity with the PayPal brand and a relatively high level of trust of PayPal.

The best place for an application is not on the carrier's portal, but rather on the device standby screen. Device user interface customization technologies such as Qualcomm's uiOne allow such access. Some carriers allow full device access; others have sharply defined what developers can and cannot do. Some carriers have also recognized the need to make applications more accessible and the user experience more manageable, and have created favorites, available from the standby screen, allowing access to any application, web site bookmark, or component of the device's user interface.

If your primary marketing channel occurs via the physical rather than virtual environment, you will not have the opportunity to display all the carrier and device rules on a poster or magazine ad; your application platform should be selected accordingly. The greatest independence of carriers is achieved with SMS or native applications; the largest number of devices supported is achieved with SMS, browser, or Java ME applications.

4.7 OTHER CONCERNS

Unfortunately, the user experience of the application itself is not the only concern in selecting a technology. Cost of deployment and access to sales channels are key marketing measures, and an organization's familiarity with a specific platform's base technology is also important. There are times when an organization needs to step out of its familiarity, but cost of deployment and access to sales channels are always relevant.

The Carry Principle dictates that devices are small and wireless, so they therefore have a limited battery life. There are three major demands on the battery beyond simple standby: screen display, network usage, and vibration.

Different application technologies draw down the battery differently. Text messaging, for limited interactions, uses very little battery. In contrast, multimedia messaging uses more both due to the larger downloads and because the user will spend more time looking at the pictures than simply reading a message.

Local applications require some processing and a lot of screen display, so they are roughly equivalent to multimedia messaging.

Web applications require both screen and connectivity, so they have higher power requirements than everything except applications using vibration.

4.8   PLATFORMS

Different platforms have different strengths and capabilities for development. Table 4.1 summarizes capabilities of some standard platforms. Keep in mind that of all the sections of the book, this is the one most sensitive to changes in technologies. Before making final technology decisions, research the most recent capabilities of a platform and monitor how much of the device market has the updated technology.

Messaging is a catalyst technology, enabling a more robust user experience for myriad applications. A voice-only application can send requested information via messaging, adding visual and local storage components to the experience. A message to a short code can return a link to an application or web site, bypassing complex URL entry while providing user identification to the server. Indeed, messaging can enhance the experience of an application built on almost any platform, if the application is built to handle it.

Applications can certainly be written with messaging alone, and the selection of text, voice, and multimedia messages gives an array of possibilities. These are asynchronous in nature, with local data stores.

Note that text messaging is essentially a command-line user interface. All reports of 'ease of use' are largely a function of access to text messaging on the phone and environmentally available help prompts. Any application with extended text messaging input needs to be carefully designed with robust input processing on the server.

Mobile browser technologies started with HDML and proceeded to the Japanese cHTML and the European and American WML. These technologies merged, in a way, to become WML 2.0, which is XHTML Mobile Profile plus extensions allowing the advanced navigation features found in HDML and early WML. Unfortunately, few browser vendors implement the navigation features, and some implement only XHTML Basic, so the de-facto standard for new development is XHTML Basic[5] – with external style sheets using a stripped-down CSS.

---

[5] XHTML Mobile Profile is XHTML Basic plus the tags <b>, <big>, <fieldset>, <optgroup>, <hr>, <i>, <small>, and <style>, the 'style' attribute, the 'start' attribute on <ol>, and the 'value' attribute on <li>

Table 4.1 Platform characteristics

Platform	Input	Interaction responsiveness	Storage	Display	Supplemental	Multi-device deployment cost
VoiceXML	Speech only	Fast	Remote	Aural	None	Low
Standard browser (XHTML, cHTML, WML, CSS)	Buttons	Slow	Remote plus cookies	Visual	Low	Low[a]
Java ME	Buttons (visual, speech sometimes possible)	Medium (varies with device)	Local plus	Visual, aural	Medium (varies)	Medium
BREW	Buttons, visual	Fast (native level)	Local plus	Visual, aural	High	Medium
Scripted browser (web 2.0, AJAX)	Buttons	Medium	Remote plus cookies	Visual	Medium (varies)	Medium[b]
SMS, MMS	Buttons, visual	Asynchronous	Transient	Visual (SMS is text only)	None	Medium (MMS); low (SMS)
Flash/Flash Lite/SVG	Buttons	Medium	Local plus	Visual	Low	Medium[c]
uiOne	Buttons	Medium	Local	Visual	High	Medium
Native (Palm, MS eMbedded C++, Symbian C++, Linux)	Buttons, visual, speech	Fast	Local	Visual, aural	Very high	High
Abstract Native (Python, OPL)	Buttons	Medium	Local plus	Visual, aural	Medium	High
3GPP, 3GPP2, media	Buttons	Slow	Remote, local	Aural, visual	None	Medium

[a] There remain enough rendering differences in devices that testing on multiple devices is desirable.
[b] Scripting capabilities are highly variable across devices.
[c] Flash requires separate compiles for different device configurations, although the same design often can be used.

Some browsers also support scripting, although this requires more processing abilities. Opera Mobile supports full AJAX, but only for a limited number of devices. Expect AJAX access to local data stores to vary almost as much as Java ME's access to local data stores. Other browsers support ECMAScript only; again, support is highly variable.

Java ME, BREW, SVG, and Flash Lite were all designed as application platforms with cross-device porting. Similarly, OPL was designed for rapid development of applications to run on myriad Symbian devices. As such these platforms abstract the capabilities of individual devices to a (mostly) common set of capabilities, and do not have access to other device capabilities. Flash Lite, for example, cannot access the volume buttons on a phone; many Java ME MIDP 1 and 2 phones have no access to volume control.

Cross-device application platforms have several implementation issues, particularly when different vendors write the application environment. Applications are supposed to work across devices, but this fact needs to be tested. It is not uncommon for the quality assurance team to be twice the size of the development team for a Java ME development organization.

Native application environments, such as Symbian C++, PalmOS, Linux, and MS eMbedded Visual C++, allow deeper access to the device capabilities than do the cross-device platforms. They run in the native operating system, rather than in an interpreted environment or virtual machine. They are faster, with greater access, but with very limited cross-device portability.

uiOne and similar technologies allow the transformation of the device's user interface, particularly the standby screen. Most such technologies merely change the graphics, font, colors, and layouts of existing functions on the phone; uiOne has been combined with BREW to give it native-level access to device development. These technologies will have limited control over the phone, either from the inherent technology or from carrier limitations.

There are a number of media play technologies, including those based on MPEG 4 and MPEG 2, Windows media, and so forth. Device support varies wildly, but translating content is relatively painless so all formats can be distributed.

5

Mobile Design Principles

There are fundamental concepts of design that apply across all design domains, but each domain interprets how these design principles apply. For example, one fundamental design concept is Fitt's Law, which states that the time to acquire a target is a function of the distance to the target and the size of the target. The further the target is away from the user's current position, the longer it takes to move to the target. The smaller the target, the more the user has to use fine muscle control and hence take more time to move.

While Fitt originally worked on control panels and studied muscle and limb movement, the basic concept has been extended to cursor movement on computer screens.

The implications of Fitt's law varies design by design, domain by domain. The size of a target is affected by input mechanism, such as direct manipulation, cursor manipulation, or scroll and select. The distance to a target is affected by display and input mechanism, such as physical controls, computer screen with mouse, serial input (scroll and select or pure keyboard input), or small screen with stylus. What follows are some examples in different domains:

- *Hardware control panels.* Group controls used together or in sequence make important controls large and centrally located.
- *Mouse-driven interfaces* (software). The 'large' controls are the edges of the screen, as they are really infinitely large in one direction. Corners are larger still. Thus frequently used items should go around the edges. The existence of a cursor gives a precise definition of 'close', so contextual menus can be truly context driven.

Designing the Mobile User Experience   Barbara Ballard
© 2007 John Wiley & Sons, Ltd

- *Mouse-driven web sites.* When a link is activated, the screen changes, possibly completely, and the edges of the screen are not accessible by the web page. Thus 'where the cursor is' is the largest target, and cultural visual scanning practices are used to place most elements. Consistency between pages helps the visual scanning process. Note: modern web development techniques allow for an interaction style more closely resembling software.
- *Stylus-driven interfaces* (small screens). The concept of 'distance' is almost meaningless, as the entire screen is smaller than the hand and there is no cursor. Thus size and predictability of location become the key issues for speed of target acquisition.
- *Scroll-and-select interfaces* (small screens). The number of key-presses to access a target is a good measure of distance, and size is reasonably represented by whether the target is currently displayed or not. As more devices display several font sizes, target size will be a combination of visibility and target size.

Note that in all but hardware control panels, the keyboard is a known distance away (short distance) but suffers the challenge of no visual display-control association (small size).

Some issues are present in the full-sized computer world, but are exacerbated in the feature phone world. For example, phone users, like personal computer users, are not power users. This can result in features for users perceive as invisible, notifications not being dismissed, applications installed in main memory without concern for memory available, and even expired applications still on the device's main screen. Further, users do not necessarily understand memory management, and may believe that simply by inserting a memory card they have more memory – even if they never move anything to the memory card.

In addition to novel interpretations of known design principles, the mobile space has several unique principles. Each will be discussed and implications discussed.

5.1 MOBILIZE, DON'T MINIATURIZE

First and foremost, simply transferring a full-sized computer applica-tion to the mobile environment almost always results in a suboptimal mobile experience. Attempting to construct an application that works the same on both platforms will reduce its quality in both places.

A full-sized computer does not have integrated cameras or reliable voice communications; a personal communications device will not have a readily usable full-sized keyboard or large screen. Desktop users are primarily interacting with the computer; mobile users may primarily be interacting with the world, both through a mobile device and in person.

Mobilizing an application means reconsidering the entire purpose of the application, not just changing display technologies or interaction nuances. How do your users' needs change when they are no longer at their desks? Does your application even have a place in the mobile environment? Or, is your application one that doesn't make sense in the full-sized computer environment?

What mobile technologies best meet your mobile users' needs? SMS? Camera? Web? Symbian? Windows Mobile? Java ME? What devices are your users using, what carriers are they using? What features and services might they want in the future? Are bar codes a relevant part of the use environment? What about bar code readers – or perhaps the camera will be sufficient? How does the user's location affect the application's understanding of the user's context? Or is the location merely a method of reducing text entry?

Indeed, this concept, to rethink what is desirable and possible for the mobile environment and to build and rebuild accordingly, is the main premise of this book.

5.1.1   The Carry Principle

Personal communication devices differ from computers in that many if not most users always carry the device with them. This has several important implications for the mobile device and service design:

- small device – users won't carry large devices
- multi-purpose – users won't carry a variety of single-purpose devices full time
- personal device – the device is not shared, and is likely to be customized
- always on, always connected – instead of being turned on only for use, PCDs are turned off only to preclude interruption for various temporary reasons
- battery-powered
- wireless – and thus inconsistent –connectivity.

5.1.2   Small Device

The most obvious implication of The Carry Principle is that the device must be small enough so that it can be readily carried. The device will not always be with the user if it is bulky or heavy. This, in turn, triggers certain design constraints.

A small device, with a small screen, can effectively display only a single window at a time, with dialog boxes and menus. The user can thus use exactly one application at a time. An interrupted application is truly interrupted unless the device returns focus to the abandoned application. The handling of interruptions varies drastically across different devices and platforms.

Most devices are good at managing incoming messages during application use but ineffective for launching other applications or calls while maintaining application status. In particular, some browsers return the user to the home page upon each launch; these browsers cause the user to lose track of what was happening before the interruption.

The interruption problem also exists for Java and other platforms. The time to launch the application can reach thirty seconds, so an exited application reduces the likelihood of continued application use.

The single-window interaction also causes challenges in accessing information outside the application. Just as the phone book needs to be available during a voice call, movie information might be useful for a chat session. Applications should provide access to any information resources that might be needed to successfully use the application.

One-Handed Operation

Although PCDs can certainly be used with two hands, they will frequently be used with one hand. Expert users can type one-handed without looking at the screen.

A stylus-driven device may also be thumb-operated. If your touch-screen application will be used on the fly, you should also support thumb operation with larger controls for certain actions. Many users will only use the stylus when interacting with the application for extended periods, or to enter text.

Users may be interested in using your application surreptitiously, such as under a table at a meeting without looking. To support this behavior, ensure that common tasks have a stable set of keystrokes to complete the task. In particular, do not insert any controls between where the cursor is (or starts) on the screen and the main task controls for the screen. Note that this also makes your application more accessible to the blind and vision impaired.

Difficult Text Entry

Even on devices with easy text entry, such as a thumb-sized QWERTY keyboard or an integrated alphabetic keypad like Fastap, text entry is more difficult than on full-sized computers. Frequent users of text messaging may type relatively quickly, but they do not do it for any length of time (text entries tend to be short) and they use shortcut abbreviations wherever possible. Intrinsically, they recognize that text entry is difficult.

Predictive text is also relatively difficult, even though it makes a hard task easier. While expert use of QWERTY keyboards and even triple tap involves focus on the screen, most letter prediction mechanisms create significant cognitive dissonance if focusing on the letters. The user can be typing one word, but the screen is displaying another because that letter combination is more frequent. This can slow down the text entry process.

In the future, full-sized QWERTY keyboards may be more common. Currently available are rollable fabric keyboards and infrared keyboards projected onto flat surfaces[1] requiring no separate accessory. These solutions work well in certain use situations, such as taking notes at a meeting, but they will not be the standard input mechanism for PCDs.

Use of any full-sized keyboard requires a surface upon which to place the device, a surface upon which to place or project the keyboard,[2] and the ability to type with both hands. The user's mobility is thereby limited to that of a laptop computer. If your application requires this degree of immobility, consider a laptop or tablet computer as your application platform.

---

[1] Note that projected keyboards provide no tactile feedback when a key is pressed and thus forces the user to watch the keyboard and not the screen. Still these have promise for certain niche users, where a keyboard projection could be 'thrown up' for use in contexts where either work demand (text quality or quantity) precludes other alternatives. A number of niche uses (medicine, higher education and the military) exist for such keyboards.
[2] Some inventors have created truly virtual keyboards, requiring no surface upon which to project. These 'keyboards' instead track finger positions. This will remain at best a niche solution to the text entry problem due to the requirements of touch typing and wearing sensors on the hands. Further, they still require a surface upon which to place the device, so there is not a significant advantage compared to the fabric or projected keyboards.

As for mobile devices, reduce text entry as much as possible. Pick lists (drop-down menus or full-screen lists) convert some tasks to cursor movement. Other input sources can include:

- Global Positioning System (GPS) or other location services eliminate the need to enter current location for services ranging from finding a local movie to directions to a day runner.
- Cameras can take pictures of bar codes or other code systems.
- Cameras can take pictures of text, including product packaging, business cards, and receipts.
- Address books or calendars can reduce input in certain classes of applications.
- Auto-completion[3] (built into some devices' general text entry mechanisms) reduces keystrokes for long words; this mechanism also can be added to individual applications.
- Image recognition of faces or objects can be very useful. Consider a camera application, on a PCD or on a standalone camera, that organizes pictures using similarity of faces or locations. All of the pictures of Betty are tagged 'person 1', which the user can rename as 'Betty'. All pictures taken at a specific restaurant, if recognizable, would be tagged as such and those in a specific time range would be tagged as a specific meal. Image recognition could also be used in a tourist direction-finding application.
- Date and time can be extracted from the PCD. The server time can also be used, but may not be in the user's current time zone. The application context will dictate which is preferred.
- Speech, processed at the server using dictation technologies or VoiceXML, can be used in a multimodal application. Many applications would benefit from adding a speech element, something that is more possible with packet data networks.

Small Screen

A small device dictates, to some extent, a small screen. Many PCDs will retain familiar LCD screens, but future devices may have a flexible rolled display that enables a larger screen.

Small screens cannot support multiple windows; the space dictates only smaller sizes of layered information be used such as drop-down

---

[3] Both Tegic/AOL's T9 text entry and Zi Corporation's eZiText suggest words from the dictionary that fit the current input.

menus, pop-up menus, and small dialog boxes. Thus the user will visually interact with only one application at a time, using only one window. An 'open in new window' link is the same as a normal link on a web page.

With mobile devices, users will have an even lower tolerance for screen rendering delays than they do on full-sized devices. Pre-fetch data whenever possible to speed information rendering, but be sure to provide a mechanism to turn this off for users who have to pay for each byte of data. Consider using a local application rather than a web application for rendering intensive services.

Small screens also prevent the user from smoothly reading large chunks of text. There are three reasons for this. First, it is easy to lose context when scrolling, as the physical and cognitive efforts of moving from page to page interfere with reading comprehension. Between-screen continuity is broken. Second, glare and pixel issues make the actual font difficult to read. Third, well-practiced text scanning behavior is not supported. Most people scan text for nouns or phrases to comprehend text, but the frequent line and page breaks coupled with the lack of negative space[4] makes this difficult to do, forcing users to read word by word rather than phrase by phrase.

Mobile content must be carefully designed for the small screen and lack of user focus. It could be argued that this one of most significant design challenges mobile designers face. It may be that we will have to rethink the page metaphor in much the same way we have to reject the personal computer as a model upon which to design mobile devices.

This problem of mobile content creation will be more complex when public-use displays become prevalent. This could create the possibility of approaching a display at home or at work and seeing the information and applications from the handheld device displayed on large displays. In this case it will become necessary to switch from a single-panel display to a multiple-window display. As of 2006 no such system exists.

5.1.3   Specialized Multi-Purpose

Users want several features; marketers, vendors and the mobile industry will want users to have even more. Some, or most, of these features

---

[4] In visual design, negative space, or white space, is the area the eye does not register. It is used to show the eye what path to follow. Small screens filled with text have little or no negative space.

are available on focused-function devices: digital cameras, iPods, televisions, GameBoys, calculators, and watches are all viable useful products. Few people are willing to keep all these devices in their pockets at all times. But the question remains how to determine which of the many functions should be implemented device by device, market by market. Answering this question requires further empirical research into mobile user needs – a task many device manufacturers have refrained from either underwriting or doing themselves.

Focused-function devices, or information appliances, are devices built around a single purpose. Other functions may be available – the iPod has a calendar – but the main experience is not sacrificed in any way. The calendar does not impinge on the music experience. Information appliances are used by people who cherish the experience of using the particular feature or service.

The Carry Principle dictates that PCDs be multi-purpose devices somewhat like computers. The PCD will first have all the features that are desirable but are not, in the user's opinion, worth carrying a separate device to experience. Further, even features that do merit an information appliance (single devices) will be included in the PCD simply because it is always with the user whereas an information appliance typically is not. If the experience of using a feature is important enough to the user to justify an information appliance, this user would likely appreciate having access to the experience at any time. There is also of course a market logic for providing mobile users with the features they typically associate with information devices.

This is not to say that there will in the foreseeable future be a stabilization of PCD design like there has been with personal computers. Different features are important to different people, and for mobile devices these features radically affect device design. A person who plays games to fill time while commuting may be content with five steps to start a game and generic phone controls; a dedicated gamer might prefer a GameBoy phone. Both devices could have the same features, but very different design.

Already popular are 'hiptop' and BlackBerry devices, which focus on text messaging. Some of these are fully functional voice phones for people who prefer text to voice communications. Form factor proliferation will continue as long as new niches are identified. In fact there is a market opportunity for vendors and service providers who can provide as much differentiation as possible regarding both devices and services.

Bluetooth and other local near-field wireless technologies have the capacity to connect devices together, which is important for users who use multiple information appliances. A separate PDA ought to be able to cause a phone to call or text a specific contact. A GPS device ought to be able to use addresses from the PCD. A device should be able to access the Internet via a local wireless connection. This capability allows an even wider array of devices to share PCD characteristics.

Thus The Carry Principle dictates that feature creep abounds, but that there will be no stabilization of design. Users would not want the same shaped device any more any more than they would all want the same type of automobile.

User Interface Styles

Devices have their own particular user interface styles, with customary use of softkeys or typical organizations and visual styles. There is no common style, due to manufacturer differentiation, manufacturer patents, and different needs with different capabilities.

A simple, low-feature scroll-and-select phone is best used with some type of rocker key and activation. Nokia-style softkeys ('Options' and 'Back' as the softkey labels) are common but do not test well with novice users; softkeys aren't even required for good design. Some phones have both an activation button ('OK') and softkeys. Regardless, the user is accustomed to her device's user interface.

Matching the device's user interface style is important to usable applications. Some markets have gravitated towards standard interface paradigms across manufacturers. In India, for example, devices tend to have a Nokia-like Options/Back softkey user interface because that is expected. The Nokia interface is thus, in India, regarded as intuitive.

Scroll and select phones with a large number of features can suffer from the default tree hierarchy paradigm breaking down. The large number of features force users to navigate deep into a complex hierarchy unless the desired feature is one of the small set that are readily accessible – and recognized as such. The industry is seeing the emergence of new methods for working with large amounts of features and content using the same interface methods. Content and features can be accessed through bookmark-like favorites. Themes allow user interface customization, pushing preferred features higher up in the hierarchy. Expect new paradigms, such as organization by frequency of use and meta data, to emerge.

Stylus driven devices have more flexibility in user interface, and use that flexibility for market share differentiation. But some users reject stylus use and find the hand–eye shifts between stylus and touch difficult to master. Windows Mobile was designed to support large quantities of features and content; Palm was designed with fewer capabilities in mind.

Each user interface has its own advantages and disadvantages, but provides the context for your application. Among other things, this means that testing your application on an arbitrary device will provide little useful data for users accustomed to a different device.

Some platforms, particularly Java ME, try to account for user interface differences by not specifying how certain features are rendered. In theory this allows the application environment to match the device user interface. In practice, device manufacturers seldom consider the impact of the application environment implementation on the user and simply do not specify how the environment should be displayed.[5]

Rendering Idiosyncrasies

Devices have different capabilities, input mechanisms, display characteristics, and user interface paradigms. Due to varying user needs, this will remain true. Thus rendering differences, and the resulting opportunity for creating competitive advantage, are a fact of life.

Further, even standardized platforms have their implementation problems. One browser developer may have decided that background images were inappropriate in the mobile space. Another may have been unable to code proper table behavior due to limited processor capabilities. One designer may have thought that both softkeys could bring up menus; another designer may have limited menus to the second softkey. These differences can exist even with devices with largely the same characteristics and largely the same user interface. This of course raises questions for users and can make devices with the same features and standard platforms seem counterintuitive.

Rendering idiosyncrasies, combined with differences in feature implementation, cause 'write once, run anywhere' to be an unfulfilled dream. We do not expect the dream to be fulfilled. See 'Handling

<hr>

[5] This statement is made based on both observation of myriad devices' implementation of Java ME's KiloByte Virtual Machine (KVM) as well as experience working with carriers and device manufacturers who did not have KVM implementation anywhere on their priority lists.

Device Proliferation' later in this chapter for some suggestions about how to manage this challenge.

5.1.4 Personal Device

A PCD is like a wallet or a purse: its loss will be noticed and rectified quickly. Its connectivity will be discontinued, and transferred to another device. The carrier may be able to remotely erase the device so the data is unavailable to anyone who acquires a lost device. These simple facts have a number of implications for the design of security in applications:

- Password entry need not be masked. The user can readily hide the screen from onlookers, more easily than hiding which keys are being pressed. Further, the difficulty of text entry makes password entry costly to the user experience. Of course, some applications do indeed need that extra security, but those are rare.
- Account cookies should not expire quickly. The fact that users will disable their network access upon device loss means that any thief cannot get to sensitive online data.
- Some sensitive data can optionally be saved on the device. If the user is known to have access to remote erasure of device, then private information can be stored there.

5.1.5 Customized Device

Ringers, wallpapers, stickers, and face plates are some of the ways users customize their PCDs. Because they are personal and visible, PCDs can become statements about the personalities and status of their bearers. The device is an accessory as much as it is a communications tool. In effect personalization (and the market advantage it offers vendors) is something that makes mobile device design a different kind of technology and marketing arena than personal computers.

Newer devices can also allow customized user interfaces, sometimes known as themes, which allow for further personalization. The increase in popularity of this technology means that even if an application knows what model device is being used, the exact environment, even user interface, is not guaranteed.

The importance that customization has for mobile users needs to be explored more. But essentially we believe the market for customization could in time rival the market of goods and services like data and data delivery in the mobile industry. Certainly the current commercial success of ring tones suggests this will be the the the case.

5.1.6 Always On, Always Connected

Society is still learning how to deal with prevalent mobile phones, with alerts to turn phones off in theaters and nasty glares when a phone user is being discourteous. There are fewer and fewer places where one can escape from mobile phone intrusion.

While public rest-room culture may not appreciate mobile phone conversations in rest-room stalls, voice phone use does exist. In all likelihood, there is significant non-voice use of PCDs happening in rest-room stalls. Theaters and churches exhort people to turn their phones off; many instead switch phones to silent or vibrate modes and they resort to text messaging.

These examples illustrate the degree to which not only is the PCD always with the person, but that it is always on. Many users often feel a kind of withdrawal when disconnected from the virtual world, whether accessed via their mobile devices or their full-sized computer. This feeling of loss is similar to what many people feel when disconnected with their television.

5.1.7 Battery-Powered

A carried device is not connected to a power source but is instead powered by batteries. In places with unreliable electricity, this actually makes carried devices more reliable than many fixed-location devices. Battery power and wireless connectivity could go a long way to equalizing the infrastructure inequalities between industrialized and lagging economies.

Although the mobile user is not tethered to an electrical cord during use, she still cannot roam far without a charger in her briefcase. She will have to reconnect at the end of the day. Batteries with large capacities are available, but their larger size makes the device heavier.

Most people will not want to charge their devices every day. Processor power, screen display, and connectivity all increase the demand for power. Size limits the supply. This power restriction means that anything the device can do to limit power use, such as dimming the display or even using powerless displays, should be considered.

Similarly, anything within reason that an application can do, such as reducing connection time, not waking the display when unnecessary, and reducing processor demands, it should do.

5.1.8 Inconsistent Connectivity

A carried device is by definition connected to information sources wirelessly, and from different locations. Wireless networks have service holes or outages. Cellular networks have dead spots, especially in the United States but also in tunnels and basements worldwide. Wi-Fi hot spots are inherently spots (and thus spotty), and further have a limited number of possible users. Even wide-area wireless networks like WiMAX will have dead spots, limited coverage, and the inability to penetrate to the middle of the mountain. Thus inconsistent connectivity is an integral part of using a PCD, especially when on the move.

Applications need to be designed to handle inconsistent connectivity gracefully. One reason users rejected early WAP implementations was due to a failure to handle this problem. Nokia browsers[6] required a live connection to the Internet to run, and when the Internet connection dropped the browser exited, even if the user was merely looking at a page and not requesting data. To make matters worse, these browsers always started at the home page when launched. Thus a user who dropped coverage for even a second while trying to accomplish some task would find all his work erased and unrecoverable.

SMS gateways handle inconsistent connectivity by resending the messages when delivery fails. This is an excellent method of handling the problem, but marketers should avoid using the term 'instant' when describing text messaging.

If your application contains infrequently changing data to which the user needs reliable access, a local application is better than a web application. Pre-fetching data, whether in a web application or a local application, will help ensure that the data is available when the user asks for it – whether the network is or not. Unfortunately most browsers today have very limited pre-fetch support.

---

[6] Nokia's chief browser competitor at the time, Openwave, encouraged calls to drop to avoid costs, and started at the last visited page to avoid loss of work. Unfortunately Openwave's feature list was not well known or understood, and Nokia's failings affected the entire industry. Usability guru Jakob Nielsen, in his company's report about WAP usability, condemned the entire WAP concept based on Nokia behaviors.

5.2 USER CONTEXT

A desktop computer user is sitting with a computer at a desk. A laptop user might have taken the computer to a coffee shop, library, airport, or meeting room... but largely will be sitting with two hands on the keyboard, the device on some surface. Mobile device contexts are more varied, and more difficult to predict and discover.

Mobile devices share with ubiquitous computing the ability to discern user context. Where mobile devices make assumptions about one user entering various situations, ubiquitous computing systems make assumptions about all people who enter a space. A ubiquitous computing device can be set up to take into account facts about the immediate environment, what information is available, and what tasks are likely, and displays information accordingly. The mobile device knows nothing about the environment but has the resources and features that could enable it to learn much about its user.

Myriad sources of information are possible, some gleaned from the environment and others intrinsic in the information on the device:

- Geographic location, such as from GPS, can determine travel status, whether the user is likely to be late for a meeting, or what the user is doing. For example, if the user's location is on a train line, the user is probably on or waiting for a train.
- Precise location, such as from a Wi-Fi network, Bluetooth, or an RFID[7] reader, can enable extremely targeted marketing or very local information transfer.
- Motion and temperature sensors within the device can detect user movement, air temperature, and gestures. These could possibly be combined to intuit mood.
- Calendars can provide likely user activity. If the user is in a meeting, sending advertising is inappropriate. However, sending industry news may be very appropriate.
- Cameras can either capture images directly, or recognize image contents such as bar codes, faces, traffic signs, or other environmental data.
- Local data sources, accessed by Bluetooth, RFID, Wi-Fi, or other mechanisms, can be used to allow the local environment to talk to

---

[7] Radio Frequency Identification tags are inexpensive chips that can have information stored on them; they can be read by nearby readers but require no power themselves. A phone could have a chip, a reader, or both.

the mobile device. A store shelf transmitter could offer a coupon for 20% off a specific product, as long as it is purchased within the next 15 minutes.

- Other personal devices can provide a wide array of information, limited only by designers' imagination. Apple's Airport Express, which can route music from iTunes to a stereo, provide a stationary example of device interaction.
- Other users' mobile devices are another source.

When these and other information sources are combined intelligently, they can give the users enormous benefits which we are beginning just to explore and exploit. Travel applications can combine several of these sources with online information to alert the user when she needs to leave for the airport, even in an unfamiliar city. If the user is out of the office and near restaurants at lunch, any place with a special or matching the user's food interests could send information to the user.

As time goes on, more sources of context will become available.

5.3 HANDLING DEVICE PROLIFERATION

Device proliferation is a reality of mobile application design. Many attempts have been made to create some platform, some technology that allows developers to write once and have the application run on every device, but none has succeeded. Sun's Java ME (itself a platform targeted at a class of devices) has itself expanded into many nonstandard implementations partially driven by devices with different capabilities. Different browsers have different capabilities, and different carriers allow different functions.

Handling device proliferation is a necessity. There are four basic approaches to designing an application to run on multiple devices:

- *Targeted* – select a set of targeted devices (mobile and full-sized) and then write an application that works on them only.
- *Least common denominator* – select technologies and designs that will work on all devices (includes graceful degradation of code such as <code><noscript></code> tags in web pages).
- *Automatic translation* – use a technology that converts some standard core function, perhaps written in XML, into the format needed by each individual device for 'optimal' design.

- *Class-based* – identify groups of devices with common use and rendering characteristics, design the core function for each class separately, and then use an automatic tool to make the necessary changes for each device.

Implementing any of the above requires some mechanism of learning device capabilities. There are technical approaches for achieving that, including open source efforts aimed at compiling device characteristics, the Wireless Universal Resource File (WURFL)[8] and J2ME Polish. No matter which approach you take, test your application on as many target devices as possible.

5.3.1 Targeted Design

The simplest approach to developing a multi-device application is to simply identify a set of devices and then develop an instance of the application for each device. This approach works well in environments where a small set of devices dominate the market, most notably corporate environments in which the device universe is known and finite.

The best way to make this approach work is to design for a platform with highly specific device characteristics, such as native Palm, Windows Mobile, or Symbian UIQ devices. This allows the application to work on future devices, thus reducing the need to update the application with each new set of devices added to the universe.

The main benefit of this approach is that the application can have the best possible user experience on each device. The drawbacks are obvious to any product manager, developer, marketer, or accountant: each new device on the market means either a large new development and support cost, or a part of the market that won't have access to the application.

If implementing this approach, be sure to account for unsupported devices accessing the application. The only thing worse than being locked out of an application is an application that acts usable long enough for significant time to be invested and then stops working on the current device.

---

[8] WURFL information is available at http://wurfl.sourceforge.net/

5.3.2   Least Common Denominator

There is much debate about whether it is theoretically possible to write one web site that will run usably on both desktop and mobile devices. The W3C[9] is advocating the 'ubiquitous web', with the argument that if mobile user agents were good enough, and site designers used appropriate design techniques, mobile devices could effectively display the same web content as full-sized (or other) devices.

As of 2006, full-sized computers have many more capabilities than do mobile devices. The Opera Mobile browser, for example, was the only mobile browser to support AJAX[10] technologies at the beginning of the year. ECMAScript is beginning to become available on phones. It is obvious that any application that wants to work on both full-sized and mobile devices has to use a subset of the capabilities available to each.

The least common denominator approach is built into the design of XHTML and Javascript: design with standard features, such as scripting, but ensure 'graceful degradation' for devices with fewer capabilities. Alternately, the application can be designed using a minimalist approach, using only the set of features and capabilities that all devices (and carriers) support, ignoring advanced and differentiating features

The problem is, neither of these approaches will work particularly well. The minimalist is worse for any large set of devices. In the web world, designs would be limited to paragraphs, lists, links, simple forms, images, and simple tables. CSS might or might not be respected; cascading CSS would not. Links or objects in tables would be unusable, the background of the page would have to be white. Automatic refresh would not be supported. Bookmarks could not be relied upon, and cookies might not be available. The user agent might not be available. If sharing code with a full-sized computer, neither email nor SMS could be used for asynchronous transport.

The graceful degradation approach is quite attractive, and has encouraged many users in the desktop world to upgrade their browsers. Mobile devices are harder, and more expensive, to upgrade, thus offering challenges to this approach. Many things can be accomplished

---

[9] World Wide Web Consortium, at http://www.w3.org/. Of particular interest is the Mobile Web Initiative, at http://www.w3.org/Mobile/.

[10] A collection of technologies that, when used in combination, allow for a more application-like web experience. Technologies include Javascript (or Java) and asynchronous XML download.

with hiding (preferably not downloading) components irrelevant to the mobile environment.

The challenges lie in three facts:

- Mobile users have varying needs.
- The mobile environment has crucial features unavailable in the desktop environment.
- The device user interface characteristics drive different architectures.

While your users may be interested in your marketing sheets, in spending half the day on your site, or in downloading large Flash presentations while they are sitting at their desks, they may not want to view this information while mobile. Further, the user is going to be in an environment where interruptions must be considered normal and frequent.

A least common denominator (but not necessarily satisfactory) approach will have the mobile user plowing through extensive irrelevant information to find the one service he needs, with frequent restarts due to environmental interruptions. The information needs to be presented in alternative forms. Similarly, desktop users have different needs: most desktop users for example will not be interested in downloading ring tones to their computer, although other sounds might be customized.

Not only should the information architecture be different for many applications, the content itself might need to be different. Further, content style varies with the medium. For example, many good verbal presentations follow the 'tell them what you are going to tell them, tell them, then tell them what you told them' formula; good journalistic writing has the 'inverted triangle' style with the big story first and revealing progressively less important details. Blog writing tends to have lots of links and tends to be written in styles like the personal essay.

The needs of full-sized and mobile device presentation are generally different. The needs of verbal and visual presentation are also different. What kinds of content and style work 'best' for mobile devices has neither been resolved nor well researched, but this book does provide some guidance on these issues.

5.3.3 Automatic Translation

Several companies, led by database vendors, have created systems that allow users to design the core logic of an application, usually in some flavor of XML, and then have the application rendered to a wide variety of devices. These engines do work, but the capabilities of the XML language can usually access a common but not universal subset of device capabilities.

This approach attacks the rendering idiosyncrasies challenge directly. To render a page on a single device, the XML file is populated with database information, the device's display language and rendering idiosyncrasies are retrieved, and the XML is then translated to the target language in the appropriate form. This is illustrated in Figure 5.1.

For thick client applications, there are development environments that mimic the above, creating executable files for each type of device without extra coding.

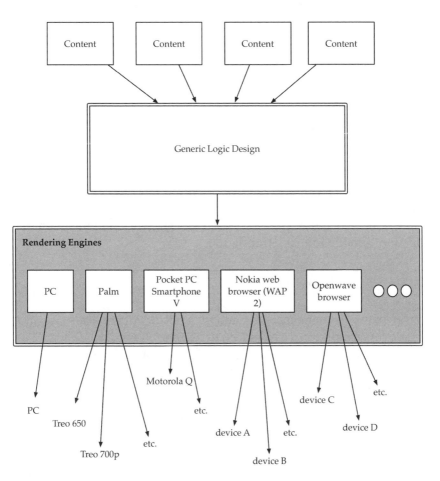

Figure 5.1   Automatic translation from a number of sources, using a single application logic, into all formats

There are dozens of these systems available, and some believe that WURFL can be used to achieve the same end. One shortcoming is that the device capabilities database rarely captures rendering idiosyncrasies, but only device capabilities. For example, two devices could have identical capabilities but might render abstract commands in MIDP in very different ways. Two browsers could be compliant with all standards but render select lists differently enough to affect widget choice. This is partially because rendering idiosyncrasies are frequently buggy and partially because the idiosyncrasies are difficult to categorize and may be unique.

The larger problem with automatic translation systems is one of architecture. Application architecture intended for a screen with seven lines of text will simply not be the same as architecture intended for a stylus-driven screen with 15 lines of text.

Use these systems systems (as described here) only in a highly controlled environment, preferably one in which users are trained to use the applications. The user experience will be suboptimal but perhaps not enough to be worth use of another approach. Continue to test on multiple devices, and do not assume that your XML application will be usable by voice or even on the desktop computer, despite any vendor claims.

It is useful to note that automatic translation systems can also be used for localization logic.

These systems can generate consistently usable designs when using class-based design, described next. Be sure to ask for the extra abstraction level demanded by class-based design when talking with the vendors about automatic translation.

5.3.4   Class-based Design

To enhance the user experience and functionality of applications, automatic translation can be enhanced to become class-based design. With this approach, identify what classes of devices your application needs to support, then write the application's core logic for each class of device. The rendering engine then does the detailed translation for each device.

Identify a set of device classes suitable to your application. How exactly to do that may depend on the needs of your application. For example, an application with a large amount of important information on the full-size version's screen will have navigation challenges when trying to map a single desktop screen onto a single or group of

Table 5.1 Potential device classes

Display modality	Device category	Relevant features
Large screen	PCD (stylus)	Camera with voice
Small screen	PCD (scroll or gesture)	Camera without voice
Large screen (voice response)	Industrial/Corporate device	Voice without camera
Small screen (voice response)	Information appliance	Neither voice nor camera
Voice only	Voice computer	

small screens. These information flows are likely to be very different for smaller screens, so device classes should be based on display capabilities.

If the application uses voice response and visual display, a stylus design will be significantly different from a scroll and select design, as there is no cursor to indicate focus. Interactions need to be different, and architecture may be different.

Classes may also be based on user interface (UI) paradigms (see Table 5.1). Applications could have information architectures in the Windows, Palm, UIQ, and BREW styles, although the core logic would be the same in each. Using the automatic translation system coupled with UI paradigms as device classes allows access to devices with copycat user interfaces, such as some Linux devices.

I have yet to find any automatic translation system that supports class-based design, but the change in system architecture is feasible. Each supported device would have to be identified as belonging in a certain category, and the device category would specify which of a small number of core logic documents to use; the revised system is illustrated in Figure 5.2. If your organization has a large enough contract, your vendor should be willing to make this change for you.

While a class-based design system does require writing and supporting more than one version of the application, the number of supported versions is small. Given the small cost involved, the user interface designers and developers can use class-specific features, application architecture matches the predominant class device user interface style, the experience feels custom to the device, and the user experience is greatly enhanced.

More details on device class selection and management can be found in 'Mobilization', in the next chapter.

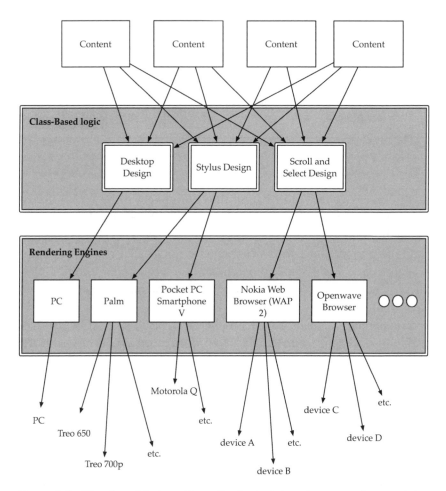

Figure 5.2   Class-based design with application logic based on group characteristics of devices

5.4   EMULATORS AND SIMULATORS

Mobile software emulators, which use the same rendering code as the mobile counterpart, are theoretically a cheap and easy way to test applications. Simulators are slightly more suspect. Despite this, developers and designers tend to rely on simulators and emulators to test their applications. They should not be surprised, however, when the application does not work the same on the actual devices. Use simulators and emulators with great caution:

- Simulator code is frequently different from the real code, so your application will behave differently.
- The underlying system architecture is not the same on mobile and full-sized devices, so your application may behave differently even with an emulator.
- Devices have myriad rendering and implementation idiosyncrasies, both feature and user interface, that are not captured in the emulator.
- Using screen buttons to operate an emulated device is very artificial, so usability test results will be biased towards avoiding their use.
- The emulator is almost certainly not the device the usability testing participant will use or is accustomed to using.
- User behavior sitting in front of a screen is very different from when holding a device, whether in a quiet office or a crowded bar.

Emulators and simulators can be used by developers for interactive debugging of logic; they can be used for usability testing to understand some components of the information architecture. Do not use them for system testing, unit testing, or user acceptance testing.

5.5   DETAILED DESIGN RECOMMENDATIONS

From the discussion in this chapter, it should be clear that detailed design recommendations have to focus on a particular platform, and perhaps also on a particular set of devices. Fortunately, there are several good sources for design recommendations, frequently called style guides.

This chapter cannot provide a comprehensive list of guidelines, but can give you some good suggestions on where to find recommendations.

5.5.1   Platform Providers

Platform providers are the most obvious source of design recommendations, as they know the development environment and design intentions intimately. In theory, if applications work well, the platform will be more widely adopted. In practice some providers do not provide a comprehensive set of recommendations.

- *Windows Mobile Design Guidelines*, from http://www.msdn. microsoft.com
- *MIDP 2.0 Style Guide for the Java 2 Platform*, Micro Edition, by Cynthia Bloch and Annette Wagner of Sun Microsystems
- *Graphical Browser Application Style Guide*, and similar documents, from http://developer.openwave.com
- *UIQ Style Guide*, from the developer and technology section of http://www.symbian.com
- *User Interface Design Guidelines*, at http://brew.qualcomm.com
- *Palm OS® User Interface Guidelines*, at http://www.palmos.com

5.5.2 Standards Organizations

Standards organizations also provide design guidelines, ones that often reflect a particular agenda. The W3C, for example, is pushing guidelines that will make applications work on both full-sized and mobile devices, which may not be ideal. The Open Mobile Alliance, in a former incarnation, provided a WAP style guide for designing 'generic' sites to run on Ericsson, Nokia, and Openwave WML 1.x browsers despite radical rendering differences. This was a least common denominator approach, and sites designed with those 'generic' rules were at best very simple.

- *Mobile Web Banner* ('WAP') *Advertising Specifications* standardizes web banners for advertising on mobile phones. These guidelines ensure consistency and adequate usability. See http://www. mmaglobal.com.
- *Mobile Web Best Practices* is the W3C's attempt at specifying how to write once and run anywhere. The guidelines are largely reasonable. See http://www.w3.org/Mobile.

5.5.3 Carriers and Device Manufacturers

Carriers have the most motivation to have useful and usable software and web sites, since these drive increased usage and revenue. Device manufacturers want users to purchase their devices a second, third, in fact many times, so a good device and purchased-software user experience is important to carriers. In our experience the carrier and manufacturer style guides are the most comprehensive for developing for the limited environment of the carrier or device type.

As devices support different platforms, each of these sources may have several guidelines. More companies and their developer programs are listed in the Companies appendix.

- Forum Nokia has an extensive technical, marketing, and design library for Java ME, Series 40, Series 60, Series 80, and web applications with separate documents for games. See http://www.forum.nokia.com
- Sprint Nextel has web, Java ME, and multimedia style guides, but some guidelines are only available if you have a partnership with the company. See http://developer.sprint.com
- Sony Ericsson has some limited guidelines for various platforms. See http://developer.sonyericsson.com
- Verizon information is found at http://www.vzwdevelopers.com/aims
- Motorola provides support for specific devices. See http://developer.motorola.com.

5.5.4 Third-Party Guidelines

Occasionally a third party, either an individual designer or a usability consultancy, will write design guidelines. Serco Usability Services may have been the first company to do this, but their WAP guidelines are neither current nor currently available. Bloggers and other online writers make design recommendations, but their recommendations tend to be rather subjective and the rationale for design choices are seldom clear or well defended. In short, online resources tend not to be very strong. There are, however, at least two exceptions to this general rule.

- Little Springs Design[11] offers style guidelines intended to cover all devices for a platform. These are available for web, Java ME MIDP 2, and media content production. See http://www.littlespringsdesign.com
- Serco Usability Services provides a varying source of guidelines in their Research section. Most of these guidelines are not connected to a specific platform. See http://www.serco.com/usability.

---

[11] Barbara Ballard is principal of Little Springs Design, which also writes many of the Sprint guideline documents.

6

Mobile User Interface Design Patterns

User interface (UI) design patterns are good solutions to standard user interface design problems. While neither standard practice nor academic research has yet formalized what a pattern is and is not, patterns have become a good method for a new user interface designer to learn good, well-practiced solutions. At a minimum, UI patterns provide a good starting point for specific parts of an application.

Clearly there is no end to the list of all possible design patterns, and a single chapter within a book is not going to describe the majority of them. Thus the patterns identified in this chapter provide more of a set of examples from which a pattern library could be built. Many of the patterns are also good examples of how mobile design is different than desktop design, or how mobile device type and user interface style influences design.

6.1 ABOUT USER INTERFACE PATTERNS

A design pattern documents known good solutions to frequently occurring design problems. In some cases, the solutions themselves become encoded as user expectations: an application that violates the common design could jar user expectations.

User interface design patterns are generally identified and articulated by design experts. They can then be used by less experienced designers or by designers wishing to create a consistency in user experience.

Designing the Mobile User Experience Barbara Ballard
© 2007 John Wiley & Sons, Ltd

If writing about 'usability patterns' is included, there are three types of UI design pattern: patterns of practice, user interface design structures, and corporate patterns. Patterns of practice are closer to best practices in development, such as processes for targeting multiple markets, and are not reflected in this book.

User interface design patterns, or 'universal patterns', are solutions that likely work across a wide range of applications and on different platforms, although some patterns are platform-specific. In addition, organizations with a complex set of offerings may also create a set of highly specific, fully stylized, 'corporate patterns' in a pattern library frequently with code associated with each pattern.

6.1.1 Mobilization

While the world of desktop design patterns all assume a consistent set of capabilities of the computer, patterns targeted at the mobile space must take into account the varying capabilities and user interface styles of the native operating system.

Some UI design patterns, particularly the aforementioned 'usability patterns', are identical to desktop design. Other patterns vary due to size of screen, cost of connectivity, input mechanism, technologies available, etc. In general, be suspicious of any desktop navigation or screen layout pattern – it may not mobilize well.

Mobile design patterns do not follow a strict categorization by application development platform. There are some portions of the wml namespace that, if present, enable interaction like AJAX or even Java ME. Thus a solution for one platform might be useful for a wildly different platform.

Using a Device Hierarchy

Desktop UI design patterns are reasonably stable regardless of platform. Tab navigation may look different in a Windows dialog box than it does on the Apple web site, but the basic concepts are the same. Only when multiple rows of tabs are needed does the underlying platform have much influence over design.

In contrast, mobile patterns rely on both device user interface style and platform. Whereas tabs are a useful mechanism on a stylus-driven device (web or local application), they are less useful on a scroll-and-select device application, and should be implemented as horizontal navigation instead. The same navigation in a web browser on a scroll-and-select device should either avoid the problem altogether, or use a drop-down list.

Since good design depends so frequently on device characteristics, a device hierarchy is helpful when working with mobile UI design. The hierarchy organizes devices into relevant device classes, with varying degrees of specificity based on the level in the tree.

There is no one correct hierarchy. Any hierarchy design will have its challenges. The Figure 6.1 sample hierarchy shows one possible organization, assumed by the designs in this chapter.

The highest node in the hierarchy, as illustrated in Figure 6.1, is a mobile device. The distinction with the most impact on UI design is scroll-and-select versus stylus devices, so that may be the second level. Within stylus-driven devices, the operating system likely has the next most impact on design decisions. Within scroll-and-select devices, softkey management paradigms may have the next most relevant impact.

Feeding into the hierarchy at the lowest level are devices themselves, as reported in a device description repository. Several of these exist, as they are included in both WURFL and J2ME Polish. The W3C envisions myriad device description repositories available.

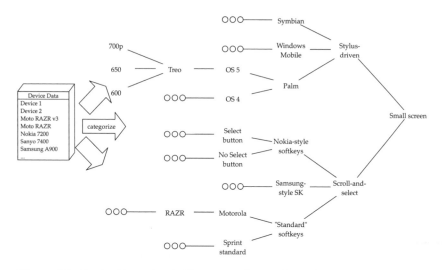

Figure 6.1 Device hierarchy, fed by content from a device description repository. Each UI design pattern applies to one or more nodes in the hierarchy. Some patterns apply to all devices. Others apply only to lower nodes in the hierarchy. Most apply to the entire hierarchy, with different versions for different nodes. In this chapter are patterns in all three categories. Figure 6.2 illustrates how patterns may apply to different nodes in the hierarchy

Building and maintaining this hierarchy cannot yet be done without human editing, as available device description repositories do not have information about user interface paradigms. The devices themselves do not report this information. Further, the hierarchy will be different for different platforms. Softkey management is largely irrelevant to a web site, very important to a Java ME or Flash application, and absolutely critical to a native application.

Each UI design pattern applies to one or more nodes in the hierarchy. Some patterns apply to all devices. Others apply only to lower nodes in the hierarchy. Most apply to the entire hierarchy, with different versions for different nodes. In this chapter are patterns in all three categories. Figure 6.2 illustrates how patterns may apply to different nodes in the hierarchy.

When designing an application, use information about target users, their devices, their training, and their diversity to help determine development strategy. Combine user and device information with project needs, application complexity, and organizational capabilities to decide what set of nodes to target.

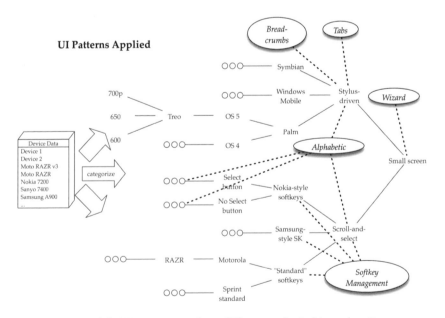

Figure 6.2   Mobile UI patterns apply to different nodes in hierarchy. One pattern can have different implementations for different nodes

A corporate intranet application might not have a large enough user population to justify multiple designs, and a generic design might be possible. In some companies, a generic scroll-and-select design might be optimal if few or no employees have PDA devices. A very simple web site is likely to work well with a generic mobile design. On the other hand, a highly interactive application or a frequently used application like a browser or email client will be well-served with a stylus version and different versions for various scroll-and-select user interface paradigms.

Designs within the hierarchy are inheritable. If targeting the Nokia-style softkey node, a design situation with no Nokia-specific design simply inherits the scroll-and-select pattern. Similarly if no scroll-and-select softkey design is present, the generic mobile pattern is inherited. In this way, targeting three nodes does not mean three times the design and coding work of targeting a single node.

The device hierarchy is an efficient, repeatable process for achieving class-based design, as discussed in Chapter 5.

Creating a Mobile UI Design Pattern from Scratch

To mobilize a current desktop design pattern, it helps to be familiar with a wide variety of mobile applications. There is a general process:

- Start from scratch for the design. Reuse the design situation, but design a new user interface.
- Decide what device classes you need for this particular pattern. It likely will be either your standard set of classes or a more generalized set, such as 'all scroll-and-select devices'.
- Consider user needs, device context, platform capabilities, and device input and display mechanisms.
- Determine whether a pattern already exists for this need. Research existing mobile UI patterns, which can be found in various places on the web, in this chapter, and in some style guides for carriers or platform providers. Modify it if necessary, particularly for different device types.
- If no mobile pattern exists, use good design and usability practices to create the pattern. Mark the pattern as untested until it has been successfully used in a variety of situations.
- Determine whether different versions of the design are needed for different nodes in your device hierarchy.
- Test. Use the pattern in various applications and test with users.

6.1.2   Universal Patterns

Universal UI design patterns can perhaps be called simple 'best practices'. They are the pure version of user interface design patterns, and apply to a wide variety of applications and across platforms. The examples in this chapter are universal mobile UI design patterns.

Most of the mobile UI design patterns found on the Internet are universal patterns. As of 2006, none of the large corporations who published their desktop UI patterns had published any mobile patterns.

6.1.3   Corporate Patterns (Library)

Many organizations, such as Yahoo!, standardize their design process using not just style guides, but a pattern library. Each pattern contains all the same information as general patterns, with the addition of specific style requirements, a concrete visual design, and frequently application code snippets.

UI pattern libraries are a logical extension of companies' icon libraries. SAS Institute, for example, makes statistical software with dozens or even hundreds of semi-independent modules; each module needs dozens of icons. Before creating a searchable, manageable library, graphic designers had to know through direct experience whether a same or similar function had an icon in a different module. Icons might conflict with each other between modules, or the same function might have different icons in different modules. Learnability and the overall user experience suffered, as many or even most users use multiple modules.

UI pattern libraries serve the same need as icon libraries, but apply to more than just icons. They also suffer many of the same challenges. Having a list of patterns or icons is insufficient: the library must be navigable with search, tags, and cross-links. Keeping the information up to date requires effort: adding and editing information must be easy and incorporated into the job description.

Despite the challenges, pattern libraries have several key benefits. Consistency of user experience eases learning, as users do not have to learn a new practice. The design pattern can be well-tested in user testing, with minor updates over time optimizing the design. The patterns help the user interface become part of the brand along with the visual design.

Developers can assign templates, sample code, or even actual code to each pattern. A notification system can alert them when a pattern with code in use has been updated, so they can in turn update the live application.

Even if developers do not code the pattern as an object for reference, they will become quite expert at implementing a pattern simply due to repeated use. Either way, development time will be accelerated, and there will be fewer bugs in the code. This will reduce testing time.

UI pattern libraries may be even more important for mobile applications than for desktop applications. Beyond the advantages offered by desktop patterns, hierarchy-dependent mobile patterns offer further advantages:

- There is insulation from a rapidly increasing set of target devices.
- There is a significant reduction in number of design decisions for a given application. Where desktop design is only one or two designs for a given situation, mobile design can contain many more due to number of target devices.
- There is higher compliance with device user interface paradigms, across applications.
- There is accelerated creation of support scripts and web information.
- There is reduced testing with regards to devices. An application built with patterns that were well-tested on devices is extremely likely to work on those same devices without failure or trouble.

6.2 SCREEN DESIGN

The screen design patterns apply to the design of individual screens in myriad situations. Many have implications for application architecture or the design of other screens.

6.2.1 List-based Layout

Mobile devices vary in their screen dimension ratios as well as size. Some have a longer horizontal dimension; others are vertical or close to square. Unless a device is QVGA or larger, the screen orientation is an important organizing principle.

Design

A web page or application screen should be designed vertically, using lists or similar mechanisms. Paragraphs wrap, spilling down the screen.

Each link should be on its own line. Form controls should be on their own line. Occasionally a pair of closely related controls can go on the same line; consider this a variant on the list theme as opposed to a horizontal layout. Almost all the example screens in this chapter use a list-based layout.

Applicable Devices and Platforms

This layout is suitable for any scroll-and-select device with a small screen taller than wide and is smaller than QVGA (240×320 pixels). Most stylus devices are large enough to support two columns.

When Used

It can be used for most non-game screens that do not serve as the main screen of an application, and almost all web pages.

Rationale

Most mobile phones are oriented vertically, with screens taller than they are wide. Horizontal layout mechanisms, like side bars, tables, and horizontally oriented control strips at best will look squished on a mobile phone. Additionally, navigating through these mechanisms on a scroll-and-select device can be confusing and unpredictable and only variably supported by devices.

6.2.2   Table-based Layout

Many devices and applications have a launch screen, with two or three columns of icons, from which major components can be started. Stylus devices in particular have these screens as application launchers; Palm has used such a screen for over a decade.

Design

A table-based layout screen is simple, with little need for softkeys or buttons. It should have a title, two or three columns of cleanly designed icons, and a label for each icon underneath it. This design is often repeated across devices and platforms; it is likely that the device currently in your possession has a launch screen with this design as one launch screen option.

On scroll-and-select devices, place the focus in the center of the layout, not the top. This reduces keypresses necessary to reach any given icon. Do not use this technique if the items do not fit on a single screen.

If at all possible, restrict the number of items to those that will fit on a single screen. If necessary and the application users can continue to see icon details, reduce icon size to make this possible. If this is not possible, consider a different design – especially for scroll-and-select devices.

Avoid using tables as layout on web sites; if a column layout is desired, use CSS.

Applicable Devices and Platforms

The table layout is particularly effective for stylus-driven devices but can be used in very limited amounts in local applications on scroll-and-select devices. Do not use a table to lay out a web page on a scroll-and-select device.

When Used

Use on launch screens, either for a device or for a frequently used application. Do not use it on a screen with frequently changing options. Consider other designs for a launch screen with more options than can be displayed simultaneously without scrolling, especially for scroll-and-select devices.

Rationale

Caution with using a table layout is a direct corollary of the reasons behind the list-based layout, combined with the need for graceful degradation on web pages in browsers that do not work well with tables.

Tables, with icons, are good for presenting more options on the screen and promoting location memory. Users know that the browser option is in the top right corner, so they can quickly tap there. Even scroll-and-select users get some benefit from position memory, but at the expense of complex scrolling control. If the list of options frequently changes, this position memory benefit disappears.

For a set of items that cannot fit on a single screen, a table layout introduces extra complexity for a scroll-and-select device. The user has to manage decision making for each item, left and right cursor movement, up and down cursor movement, and page scrolling. This extra complexity can make the task of activating an item too complex.

6.2.3 Location Selection

This is a generic interface to obtain, save, and manage the user's location across a variety of applications and device capabilities.

Design

When needing user location, provide a screen that enables a number of methods to set it, not just automatic location or postal code entry. As illustrated in Figure 6.3, the complete set includes:

- *Home*, which can be the user's postal code entered during registration or can be empty until first used. Provide a 'Change' page for the rare case when the user's home region changes. May not be useful for travel applications.
- *Favorites*, which should be an automatically generated list of locations, with the most frequently used locations at the top of the list. Especially useful when use is likely to center around known locations; may even be useful for travel applications.
- *Find Me*, which activates the device's location fetching API. Do not include if automatic location detection is not supported.
- *Elsewhere*, allowing the user to enter postal code, city, or address. When necessary, take the user to a disambiguation page to clarify input. Do not require the user to enter any more than necessary.
- *Name location*, applied to any 'Find Me' or 'Elsewhere' location specified. The screen also needs 'Save' and 'Cancel' softkeys, buttons, or links.

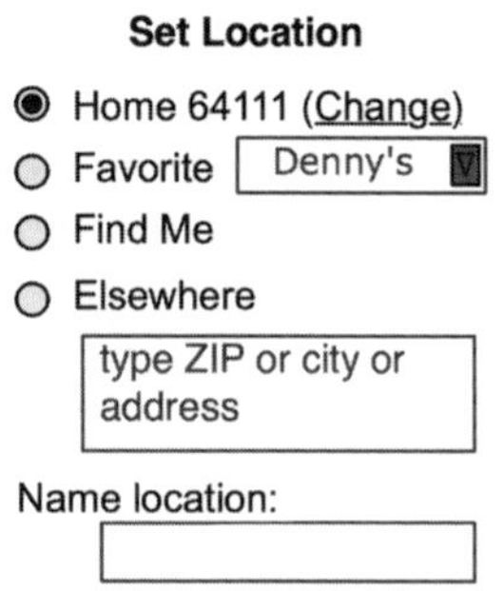

Set Location

⦿ Home 64111 (Change)
○ Favorite [Denny's ▾]
○ Find Me
○ Elsewhere

[type ZIP or city or
address]

Name location:

[   ]

Figure 6.3 Controls to set location

Applicable Devices and Platforms

It is suitable for all mobile devices and applications.

When Used

Use when location is needed at discrete, infrequent points in time. For applications that need frequent or continuous updates, rely on automatic location only.

Rationale

Devices, and plans, have varying ability to use location services on the phone. Indeed, some users may have location turned off due to privacy concerns. This should not prevent many location-enabled applications to be useful on all devices.

This design gives the user a variety of methods for entering location data, and saves data for further use without getting in the user's way.

6.2.4 Returned Results

Designers frequently ask, 'How many items should be displayed on the screen?' For older browsers, the maximum page size limited things closely. Many current browsers display entire desktop web pages on a small screen, but the user experience is less than enjoyable for many sites. What is the balance between scrolling for more results and fetching them?

See also 'Alphabetic Listings', both short and long.

Design

The optimal length for the number of items displayed from a longer list depends especially on the time to refresh the list.

Platform or node	Implementation	Rationale
Web pages (no scripting)	Limit the list to a maximum of two to three screens of results based on the current distribution of screen sizes. Place controls to navigate to the next screen at the bottom of the list.	A network request introduces a delay.
Applications	Display exactly one screen worth of results, so no scrolling is necessary. Provide navigation to the next and previous screens as controls at the top of the screen.	With no delay in fetching the next page, keeping the display limited to just the current screen eliminates the need for scrolling at all. Next and Previous controls at the top of the screen allow quick navigation.
Web pages (scripting)	Use scripting (AJAX) to achieve the same user experience as applications, above. Keep both the next and previous set of results in memory, displayable with no delay.	Same as applications.

If providing numbered access to pages, like Google results, provide those numbers at the bottom of the page. They are less frequently used and their presence at the top of the page would cause extra scrolling.

Applicable Devices and Platforms

All devices and platforms are suitable.

When Used

Use for list display whenever an application returns a list of results, unless the results naturally are alphabetic.

Rationale

There are two relevant costs to the user associated with navigating lists: scrolling through lists, and navigating between pages. The Next/Previous method of navigating between pages is well understood amongst Internet users, so the cognitive cost of using it is quite low. If the Next button has focus when the screen is drawn (either by it being the first control or by manipulating focus, depending on the platform), then a single keypress will get the user to the next page.

If there is a fetch delay, then scrolling will have some advantages over many fetches. The list is limited to approximately the size that will avoid the user being lost on a very long page.

6.2.5 Menus

A menu is a list of commands. It can be the main screen of an application, or a set of commands applicable to an item or part of the application.

Design

If the number of actions available for a given screen exceeds ten, divide the list into frequent and infrequent commands, where the number of frequent commands is eight or fewer. Provide numbered access to the frequent commands, and unnumbered, or even submenu, access to the infrequent commands. Figure 6.4 illustrates a mix of frequent and infrequent commands.

If a command is used in multiple places across the application, and is frequently used, keep both the label and the number the same throughout. This policy increases learnability for the entire application. Figure 6.5 illustrates common commands with the same numbers, even though the numbers are not consecutive.

Limit the number of commands listed on a page to roughly fifteen. Keep frequent commands clustered together at the top of the list.

Exception: if the device has an alphabetic keyboard and the platform supports letter input, construct the menu with appropriate alphabetic shortcuts instead. Limit the list to the number of items that can reasonably be displayed and mapped to letters. Any additional items should be relegated to 'More', 'Other', or the equivalent.

```
shareholder lawsuit
settlement
 Also

more (42 pages left)

Next conversation

1 Reply
2 Reply to all
3 Forward
4 Move to Inbox
5 Mark unread
6 Add star
7 Trash message

8 Compose Mail
0 Inbox
   Contacts
```

Figure 6.4 Common commands available for a Gmail message are numbered; less frequent items are unadorned links

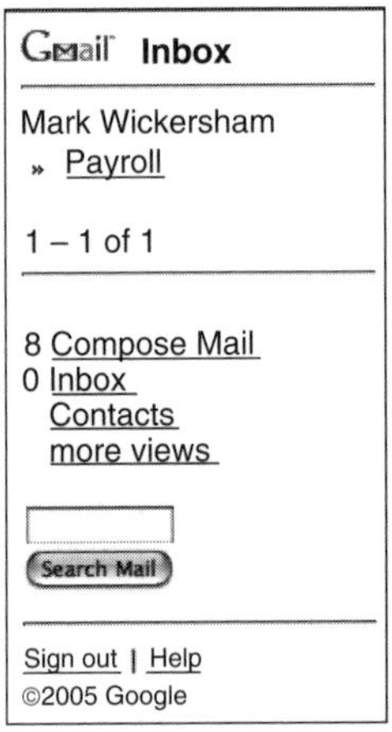

```
Gmail Inbox

Mark Wickersham
 »  Payroll

1 – 1 of 1

8 Compose Mail
0 Inbox
   Contacts
   more views

[          ]
( Search Mail )

Sign out | Help
©2005 Google
```

Figure 6.5 Gmail commands replicated in the inbox have the same number, even though the numbers are not consecutive

Applicable Devices and Platforms

Use this design on all scroll-and-select devices with platforms that support button input for navigation.

When Used

Use where the user may want to build expertise, navigating quickly using numbers rather than scrolling.

Numbered access to commands applies to any application using a page model rather than a screen model, in which vertical scrolling is assumed. This includes most list-based applications. Numbered access can be used on a non-scrolling application, but the incremental value of the numbers is lower.

Rationale

Keypresses should be kept to a minimum for common actions. Unlike on a desktop, a keypress is not simply a mouse click, but the number of times the cursor has to be moved to get to a command, then the command itself. For a Gmail message, for example, getting to 'Archive' or 'Next Message' can be ten or more keypresses. Numbered access allows that to be one keypress, although it is restricted to users who choose to learn more about the application. On the other hand, numbers do not harm usability by novices and indeed provide visual cues that certain commands are somehow different.

Keeping items clustered based on frequency is a standard heuristic for screen and control panel layout inherited from human factors. It restricts the area users have to scan for the most common items. Structure within the frequent commands can reduce scan time further.

6.2.6 Tab Navigation

Tabs are a common mechanism used to arrange more controls than can fit on a single page. Common desktop examples include Windows preferences dialog boxes and Apple.com or Amazon.com web sites.

Design

There are no changes from desktop tab navigation: what works on the desktop works on mobile, if on appropriate devices. Restrict the number of tabs to that which will fit in one row on the screen.

Applicable Devices and Platforms

It is suitable for stylus devices. Tabs are also acceptable when *all of the following apply*:

- a scroll and select device
- with four-way navigation (including left and right)

- a platform with access to left and right controls
- a platform that allows vertical scrolling to go line-to-line and not just control-to-control
- a platform with focus control
- initial focus is placed below the tabs.

None of the major browsers support all of the above. MIDP 1 doesn't support it. MIDP 2 can, but will have to be designed very carefully and tested on all devices.

When Used

Use in the same situations as desktop tab navigation.

Rationale

Same as for desktop tab navigation. The limitations on scroll-and-select devices arises from the small width and the need to scroll past each of the tabs individually. The experience can be replicated on many desktop sites: try using a site with only tab, shift-tab, letters, numbers, and Enter/Return. No mousing allowed. Tabs become quite tedious, as do left-side navigation.

6.2.7 Breadcrumbs

Breadcrumbs are a popular mechanism for locating the user within a site and providing supplemental navigation.

Design

Breadcrumb design for stylus devices is similar to breadcrumb design for desktop devices. Use the same rules of thumb for font size and color, with the caveat that some mobile devices support only one font. Ensure that the breadcrumbs are meaningful, and enabled as links. Consider restricting the breadcrumbs to only one line, with a link like '<<<' on the left edge for access higher in the hierarchy. The right side of Figure 6.6 illustrates breadcrumbs for a stylus device.

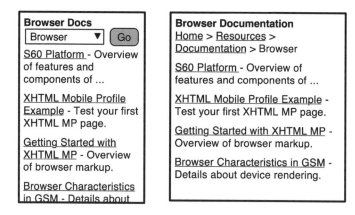

Figure 6.6 Breadcrumb navigation on scroll-and-select and stylus devices

The traditional style of breadcrumbs is not appropriate for scroll and select devices. To provide the orientation and navigation capabilities offered by breadcrumbs, use a drop-down list near the top of the screen as illustrated on the left side of Figure 6.6. If scripting is available, simply go to the relevant page when an item in the drop-down is selected. Otherwise, include a Go button. If at all possible, set the default focus below the breadcrumb controls. Note that this implementation shares much in common with computer folder navigation mechanisms.

Applicable Devices and Platforms

Breadcrumbs are most popular on web sites but can be used in applications as well. The enhanced navigation capabilities available to applications make breadcrumbs less important on non-web platforms.

When Used

Breadcrumbs are most typically used in complex, hierarchically arranged web sites. It can be used elsewhere, but the breadcrumbs tend to designate locations in the hierarchy.

Rationale

Breadcrumbs have a lot of text. 'Home > Products > Printers > HP 6131 > Specs' can have a smaller font and be unobtrusive on a desktop

site; on a mobile phone with only one or two font sizes, that information will take two or three of a device's 8 to 15 available lines of
text, and is further in the most valuable place on the page. The space
is much better used for content, not supplemental navigation.

6.3   APPLICATION NAVIGATION

Application navigation is the first place where developers from the
desktop world get in trouble. Even if connectivity, cursor navigation, and
text input were the same as for the desktop, the small screen means that
less information is displayed on each page or screen. If all the information
from the desktop screen is valuable for the mobile device, then the extra
information has to go somewhere, typically another page or perhaps the
same page accessible through large amounts of scrolling.

These patterns address some of the common navigation issues faced
by mobile applications.

6.3.1   List Navigation

Navigating between items in a list quickly causes navigation challenges.

Design

When viewing an item from a list, provide navigation both back to the
list and to the next and previous item. In this case, an 'item' can be an
individual story or picture, or it can be a subset of list of results.

There are three commands for each design: 'Next', which takes the
user to the next item or page of results; 'Previous', which takes the
user to the previous item or page of results; and 'Done' which returns
the user to the screen that called the first item.

Platform or Node	Implementation	Rationale
Web page	Below the title but above the content, place links labeled (in text or graphically) 'Next' and 'Done'. Rely on the device's back button to achieve Previous.	With no access to the softkeys, the best solution involves on-screen controls.

Application: stylus	If content is less than one screen tall, put 'Previous', 'Next', and 'Done' buttons below the content. Otherwise, the buttons should be below the title but above the content.	Control buttons on stylus devices tend to go below the content, and the decision regarding whether the next page is needed tends to happen after scanning the content.
Application: Nokia-style softkeys	'Next' is the first item in the Options menu. 'Previous' is the second. 'Done' is the third.	This is the cleanest design with Next within two clicks. Keeping Next and Previous together helps predictability.
Application: other softkeys	Make the left softkey be 'Next' and 'Done' to the right softkey. Map the Back button, if available, to 'Previous'. If not, use an on-screen command to access Previous.	Softkeys and the back button will be the fastest method of accessing these high-frequency controls.

On platforms that support it, 'Done' should also remove items from the history. Thus if the user goes to a Flickr.com album from a friend's home page, views some pictures, returns to the album overview screen, and then presses Back, she will return to her friend's home page rather than the most recently viewed image.

Applicable Devices and Platforms

It is suitable for any mobile platform, but implementation varies based on platform and input method

When Used

Use with lists within categories, in which users may not understand the categories as well as you do.

Rationale

Generally speaking, there are two behaviors when viewing a set of items, whether they are news stories, pictures, email messages, or anything that may find the user viewing more than one of the set. Users exhibiting the first behavior pick an item from the list, view the item, and return to the list. With the other behavior, users view an item, and navigate directly to the next or previous item from there.

Some users prefer the pick-and-view method all the time; others almost always prefer the view-in-sequence method. Other users will switch based on device, context, or information. In most situations, half of the users will choose one method, and half will choose the other.

Some of the recommended designs above recommended relegating 'Previous' to the Back button, even though the two commands are not the same. This decision is a tradeoff between providing the extra functionality of a Previous function with the extra user interface complexity of providing a Previous function. Since most of the time the Back function achieves the same goal, the simpler design has been chosen.

6.3.2   Game Navigation

While games vary greatly, the navigation structure to support the game should not.

Design

The basic navigation for games is fairly standard, but screen design varies with device and game. Figure 6.7 illustrates key elements of the most commonly used game architecture.

Upon launch, a splash screen identifies the application and developer. Typically next the game displays a main menu, with a first item, already highlighted, to play. Other actions such as Options (always including volume on/off and vibrate on/off), High Scores, Instructions, and Exit are in the main menu. If a game has been saved, the application displays the in-game Paused menu.

Within the play of the game itself, there must be a quick Pause function. This is frequently the right softkey but can be any number of things. When the game is paused, the 'paused menu' is displayed, which allows the user some context-specific functions including Exit. The first item in this menu is Continue, allowing immediate one-click return to the game.

When the device interrupts the user with an incoming call or message, the game should automatically pause itself.

Other screens within the game should follow best practice design.

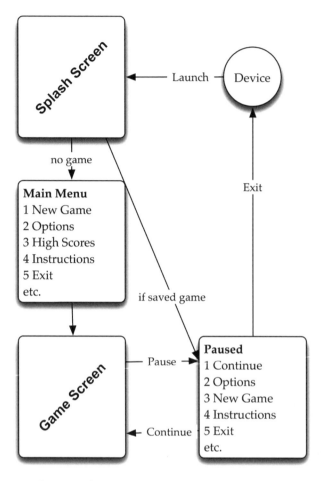

Figure 6.7 Architecture for the non-play portions of a game. Representations of Options, High Scores, and Instructions have been removed from the diagram but should remain in the actual applications

Applicable Devices and Platforms

The game pattern is suitable for downloaded applications and some scripted browser web pages.

When Used

Even games that need not have a pause function due to a lack of time pressure can use this structure. Be sure to not count pause time in any game timers.

Rationale

This design is strong due to extensive standardization across games, the need for the device to do something when paused, the need for certain navigation functions during game play, and the standard approach of starting with the main menu of an application.

6.3.3   Alphabetic Listings – Short

Any long list can be tedious to navigate. A list of alphabetically sorted items can be navigated more quickly using a combination of numbered links and fetching.

See 'Alphabetic Listings – Long' for managing lists in excess of approximately 200 items.

Design

If the list of items is two pages long (each at most three screens full of items), then simply display the results immediately. If the list of items is longer, add an intermediary level of navigation.

Platform or node	Implementation	Rationale
Scroll-and-select – standard keypad (platform supports accesskey or number access)	Group letters based on their arrangement on the keypad. Typically this is 2 = ABC, 3 = DEF, 4 = GHI, and so forth. Consider adding the number of items in each category, in parentheses: 4 GHI (5 items) Return all items starting with those letters. If desired, use accesskey 1 to directly type all or part of the name.	Accesskey provides fast activation of a link, and this distribution of letters is well-learned and readily recognized. The user need not carefully evaluate each group to determine whether the desired letter is in it, but instead just 'types the first letter' of the name.
Stylus	Display every letter of the alphabet on the screen, with two or more spaces between them. Any letter that begins items in the list is a link and visually distinguished, usually by both bold and underlining. If possible, use a larger font than usual.	Standard best practices for mouse lists. Making a letter a link when there are no results behind it destroys predictability. Visually distinguishing link and non-link letters enhanced predictability. A larger font increases the target size for the stylus, for faster more accurate access.

Scroll-and-select – full alphabetic keypad such as QWERTY or Fastap	Same as 'Alphabetic Lists – Long', below: a text entry field inviting the user to type a few letters of the item, then display matching results.	Most web browsers do not support alphabetic accesskeys and typing letters is very easy.
Web page or platform with no number button shortcuts	Same as the standard design, without number access. Alternately, use the 'Alphabetic Lists – Long' pattern.	This design is largely for simplicity: rearranging the letters for even distribution of results for a small number of devices means that each device and list result will vary, reducing both within-device and cross-device predictability.

Applicable Devices and Platforms

It is suitable for all devices and platforms.

When Used

Use when items lend themselves to an alphabetic collection but the list is not very long.

Rationale

Displaying a list of results, when the list is of manageable size, reduces the need for direct text entry, reduces issues with not knowing the spelling of an item, and also provides the user information about similar results.

6.3.4 Alphabetic Listings – Long

Sometimes the list of results is quite long, and the previously described design will not work.

Design

Provide a text entry box allowing the user to type a few letters in the item's name. Return all items starting with the typed letters, followed by all items with that string within the name. If possible, return list results while the user is typing.

Applicable Devices and Platforms

It is suitable for all devices and platforms.

When Used

Use when there are 200 or more items in the list, or when a list has entries clustered on a few letters. For example, there are hundreds of cities in California, a very large portion of which start with 'San', like San Jose, Santa Clara, San Ramon, and so forth. Even the list of US states has 19 entries starting with M, N, and O, which are the letters on the 6 button, while Kansas, Kentucky, and Louisiana are the only states on the 5 button. In these cases, consider direct text input.

Rationale

Long lists require many button presses, many fetches, and are generally tedious. In contrast, entering three or four letters to search within the list is at worst twelve keypresses and likely only five or six. States in particular can be accessed with their postal two-letter abbreviation. This is likely faster than displaying a list of items starting with a letter.

6.3.5   Softkey and Button Management

The native behavior of softkeys varies broadly across devices. Indeed, the second level of the sample hierarchy, just below scroll-and-select, addresses the native treatment of softkeys.

Design

Where possible, match application softkey behavior with native softkey behavior. For Java ME applications, consider using abstract commands rather than direct control of softkey presentation to simultaneously better match native user interface and have fewer design decisions.

Of course, consistency throughout the application is equally important. Use standard interaction design practices throughout.

Platform or node	Implementation
Nokia-style softkeys	Make the left softkey be 'Options'. All actions available to the currently highlighted item as well as the entire screen should be items within the Options menu. The right softkey should be 'Back', 'Cancel', or 'Quit', depending on context. Assume that the use of the right softkey will be overridden during text entry. In certain very simple wizard-like applications, the left softkey can be forward/select, and the right softkey can be back/quit. These applications are rare. If the device and platform has access to an End key, assign the Exit action to that key. Also provide exit functions from the application's main menu and major places within the application. If the device has no End key, add Exit to the list of controls to be allocated between the Options menu and the screen.
Samsung-style semi-softkeys (left button hardware labeled 'OK', right hardware labeled 'Menu')	Assign the most common backward navigation function to the Back button. Assign the most common action associated with individual items, especially 'View' or equivalent, to the left softkey. Assign all other controls either to a combination of the right softkey (labeled 'Menu') and on-screen objects. If the device and platform has access to an End key, assign the Exit action to that key. Also provide exit functions from the application's main menu and major places within the application. If the device has no End key, add Exit to the list of controls to be allocated between the right softkey and the screen.
'Standard' (undedicated) softkeys with Select and Back buttons	Assign the most common backward navigation function to the Back button. Assign the most common action associated with individual items, especially 'View' or equivalent, to the Select button. Assign the most common item for the entire screen, such as 'Save', 'Assign', 'Next', or similar, to the left (primary) softkey. If no screen item is available, instead choose a secondary item-based action. Assign all other controls either to a combination of the right softkey (labeled 'Menu') and on-screen objects. If the device and platform has access to an End key, assign the Exit action to that key. Also provide exit functions from the application's main menu and major places within the application. If the device has no End key, add Exit to the list of controls to be allocated between the right softkey and the screen.
'Standard' (undedicated) softkeys with Select and Back buttons . . . and with Talk/Send on the right side of the device.	Same as Standard, but have the primary softkey be the right softkey. Devices that have made the decision to put the Send key on the right and the End key on the left generally have all backward navigation on the left and forward or primary navigation on the right.

Applicable Devices and Platforms

Softkey management is necessary for any environment with control over softkeys. The most notable exceptions are web browsers and text messages.

When Used

Softkey management is a key component of the information architecture and interaction design of an application. Its strategy should be an early decision, and exact use will be a decision made for each screen and perhaps each selectable item.

Rationale

Certainly softkeys provide a dynamic label for an action that can be readily viewed by the user. While it may seem that this suggests that users can quickly learn the action of the softkeys, evidence does not show this to be true.

A user accustomed to a Nokia device, with Back and Cancel assigned to the right softkey, may never find the hardware Back button. Indeed, some applications have been reviewed and criticized for having no back function. A user of a non-Nokia device might be able to find the Back function on the softkey, but is likely to attempt to use the hardware Back button several times even after the application is well learned.

A user with a device with a dedicated Select button will almost certainly have ongoing issues with a Select function applied to a softkey. The press–OK behavior is deeply ingrained.

The standard Nokia implementation of the left softkey being used as 'Options' along with the right softkey as 'Back' can be mapped onto any device with two or more softkeys. Unfortunately this interface makes no sense to non-Nokia users. Users may think that 'Options' refers to application Options. The interface simply does not appear to have any function.

User interfaces that presume separate Back and Select buttons, on the other hand, may not work at all on Nokia devices.

While there are similar differences between Samsung and Motorola RAZR user interfaces, more designers and developers make the Nokia assumption than any other. For example, J2ME Polish prior to version 2.0 only supported a Nokia user interface or explicit softkey management.

6.4   APPLICATION MANAGEMENT

Application management patterns do not involve direct user interface issues, but instead involve less visible components of the user experience.

6.4.1   Application Download

Application acquisition is one of the first touchpoints the application has with the customer. If this is not handled through a carrier or third-party aggregator, the following measures should be implemented.

Design

From a desktop web site, the fastest method to get an application onto a personal communications device is to send the URL to the device using SMS. Ideally the URL will have encoded any personal information the user entered on the web site to be pre-entered in the application. Unfortunately, some carriers have disabled the user's ability to click on a URL found in a text message. For example, the Motorola RAZR on Verizon's network in 2005 and 2006 blocked the capability.

WAP Push accomplishes the same goals as sending SMS with a URL, but with some advantages and disadvantages. These messages are designed to not be forwardable, which protects a bit from piracy. They also may not cost the user, depending on carrier settings. Carriers that block URLs from SMS messages likely allow them from WAP Push messages. However, users must have WAP Push (or 'Service messages') enabled with the phone user interface, which may not be true by default.

From a mobile web site, a simple link to download the application is sufficient.

Applicable Devices and Platforms

These practices apply to downloaded applications.

When Used

These practices apply to applications available off the carrier's deck.

Rationale

URL entry is difficult and tedious. Many users cannot find the URL entry mechanism on their mobile browser.[1]

6.4.2 Application State Management

The application state includes what screen is being displayed, what data the user has entered, and any user settings.

Design

Good state management involves four practices:

- Save all user input except passwords.
- Discard task-related input only after the task is complete.
- Save application state, including which screen is being displayed.
- When re-entering the application, return the user to that state if appropriate. It might not be appropriate if the user was viewing transient data or if the application has not been used for a few days.

Applicable Devices and Platforms

State management is for all devices, any application platform or web.

---

[1] This includes the most sophisticated users. Journalists and bloggers have repeatedly accused certain carriers of having a 'walled garden', with access only to approved sites, even though free URL entry is allowed and a search engine allows access to both mobile optimized and all sites.

When Used

Application state management should be considered for any application.

Rationale

The user, and hence the application, is readily interruptible. The application can be interrupted at any time, by real life people, an incoming call, or a coverage hole. Thus an exited application does not indicate the user's intent to end a task.

6.4.3   Launch Process

The simple act of launching an application is often mishandled, causing the user extra delay and sometimes launch failures.

Design

This is not one pattern but instead a set of best practices.

- Check license status only when necessary. If the application is licensed for a month, check license status a few days before the old license expires.
- If frequent license checking is necessary, consider allowing a certain number of runs when no network connection is available. This allows application use in the basement and other low signal areas.
- Avoid setup questions except the first time the application is run. For example, if your application supports a 'game lobby' and the user has declined joining it, avoid asking the same question each time the application launches.
- When possible, break the application into modules. Load only the base module upon launch; load other modules in the background while the user interacts with the basic module.
- Maintain password information as long as reasonable given the security needs of the application.
- Certify the application, so the user does not have to handle queries about potentially unsafe content on a phone.

- Intelligently save context, so the user does not have to find her place again. Some applications need to start at a home screen; most applications are better off starting where the user left off.
- Provide frequent task actions on the main page. For applications with very frequent main tasks or views, the primary task or view should be the main screen. So-called 'main menu' items can go in a softkey menu or similar.

Applicable Devices and Platforms

Apply these practices to downloaded applications.

When Used

All downloaded applications need to be optimized for fast launch.

Rationale

Users tend to want to get their content, including download and purchase if relevant, within 20 seconds; some data suggests that the impatience limit is actually below 10 seconds.

On many devices, if the user has launched an application, she can do nothing else with the device until the application has exited. Only one application can be running. Thus the 30 seconds that many applications take just to launch leaves less than no time available for fetching information before the user's patience has been tested.

The promise of mobile data and applications is information and entertainment on the fly. This realization will never happen with long launch processes.

6.4.4 Cookies

Cookies are a popular method of identifying users and storing key data locally. Unfortunately cookie support varies across devices and carriers.

Design

Determine whether each cookie's function can be fully or partially accomplished through the techniques below, or other techniques. If a large portion of the site has an unacceptable user experience after reducing cookie use to its minimum, then perform a cookie test on all possible site entry pages. If the cookie cannot be read on the next page, advise the user of the problem. Most users can download a browser to their phone; Opera Mini runs on all Java ME devices and supports cookies well.

One simple technique is to add user identification data to the URL string and then having the user bookmark the URL string with ID. If worried about users sharing the identity-specific URL, add function to the site for the user to share the site easily; this will prevent users from being interested in the extra steps necessary to copy and paste URLs into other applications.

Password information obviously should not be encoded in a URL, but many applications only need password verification for a small subset of their application. Delaying the demand for the password, then allowing that user access to password-protected information for a short time as determined by your server, can bypass much of the password problem.

Applicable Devices and Platforms

Cookie management applies to browsers.

When Used

Use for web applications when the universe of browsers is not controlled or otherwise unknown.

Rationale

Some users may have cookies disabled. Other users may have cookies enabled, but their carrier or device may expunge cookies. Users who have to enter a user name and password two to three times per session of using email will quickly stop using the service.

6.5 ADVERTISING

While some applications can be funded using payment schemes like premium SMS, PayPal, or even simple applications, others need to support advertising. Advertising is a sensitive topic, as users may have to pay money for the privilege of downloading an advertisement.

Please see the appendix 'Opt In and Opt Out' for test message marketing campaign concerns.

6.5.1 Interstitials

Interstitial advertisements appear between screens. They are, in essence, a dialog box painted on top of the content of one screen before the replacement screen is displayed. They are excellent for branding. Click-through rates will tend to be low, since the user is being interrupted while attempting to achieve a goal, but studies show that retention is high.

Design

Implementation of an interstitial ad varies with platform capabilities. An ad usually has some sort of call to action, usually in the form of a link to a web site. Ensure that the advertising site is as well-designed as your application, or all advertising will fail.

Interstitials should be used sparingly. Display an ad only the first time the user accesses a piece of content, not every time. Display an ad between every four or five messages, photos, or news stories, not between every pair.

Platform or node	Implementation	Rationale
Web: no scripting	Avoid interstitials.	An interstitial ad must be displayed on a separate page, with separate HTTP/WAP requests and significant rendering. This can introduce a delay of half a minute or more. Use a banner instead.
Web – with scripting	Create page with ad image embedded, using approximately 60 % of the screen. Focus is on the image.	Mobile browsers have only one window available, so a pop-up window doesn't work. Using CSS allows

	When the user clicks the image, the associated link is followed. When the user scrolls off the image, set the CSS style for the ad to hidden using scripting.	in line display of the ad with the ability to dismiss it quickly, without any additional requests to the network.
	At the bottom of all the advertising site web pages should be a link to return to the original, sponsored, site.	The image should provide enough of the surrounding page to ensure that users know it is a 'pop-up' ad and not visiting the wrong page.
Applications: Nokia-style softkeys or stylus	When the user clicks on something that triggers an ad, display the next screen with the ad in a floating window. Include a dismiss button and a Link button in the ad, as well as commands within the Options menu.	Buttons on the screen can be easier to access than commands in the options menu if the device has a Select button. Replicating the commands in the Options menu supports both mental models and both device implementations.
	Display the ad as a dialog box, with contrasting color from the balance of the screen. Include two buttons in the dialog box: 'Skip' and <calltoaction>.	The main goal is to interfere with the user's task as little as possible while ensuring advertising content is viewed. Prominent 'Skip' controls enable this.
	'Skip' takes the user to the originally intended content. <calltoaction> can vary based on ad, with 'Link' a good default.	The cost of visiting the ad should be as low as possible without interfering with the main goal. Buttons provide rapid ad link availability, with some customization to engage in more of a dialog.
	If the user takes no action within five seconds, dismiss the ad and proceed to the application content.	
	Save the application's state so the user can return to the same point with no loss of information or navigation.	Saving the application state helps ensure the user is not punished for allowing the ad to take over the only window on the device.
	Dismiss the ad in 5 seconds if the user has not interacted with the application during that time.	
Application – standard softkeys	The same as the Nokia implementation, but with no commands in the Options menu and no buttons.	Softkeys are very quickly accessed and require no scrolling or cursor manipulation.
	Instead, the left softkey should be labeled 'Skip' and the right 'Link' or something similar.	
MMS	Require user to respond with single letter or word, or simply send content message five to ten seconds later.	

The ad content itself should follow standard best practices. It should engage the user and be relevant to the market. The amount of information should be limited to roughly what a highway billboard would support.

Applicable Devices and Platforms

It is suitable for the web with scripting or any application platform.

When Used

Interstitials are good for branding purposes, such as 'Sponsored by Coke', as well as advertising with minor interactivity.

Avoid using interstitials when the user is paying high per-megabyte data charges. This can be inferred based on type of application: if the application already uses relatively high bandwidth, a few extra images will not be relevant. On the other hand, an application that synchronizes a few data fields but largely works offline may be used by people with costly data plans.

Rationale

Well-designed interstitials give a good view of the advertising content, and get out of the user's way quickly. This is a compromise between both.

By restricting the frequency of advertisement, the user cannot simply take the habit of a second keypress to get to content quickly, as that keypress will do something different depending on whether an ad or actual content is displayed.

6.5.2   Fisheye Ads

Many applications have lists of content, including email messages, local listings, and news stories. Scroll-and-select devices facilitate changes to the visual presentation based on cursor position, much like rollovers.

A 'fisheye' ad appears with focus (highlighting) on a phone, generally within a list. The term 'fisheye' comes from a type of camera lens that works similar to the human eye, with more detail in the center of the image and less detail in the edges. A fisheye user interface has the items in focus grow and items out of focus shrink, possibly having the size be an inverse function of distance from focus. Thus a true fisheye interface

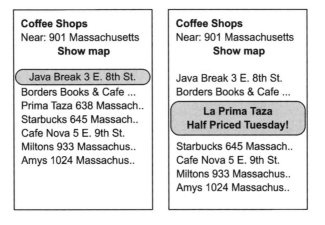

Figure 6.8 A fisheye ad is a space-efficient method of highlighting premium results. Premium listings get extra space when highlighted, including special formatting and optional graphics

for a list would have one item in the middle of the screen, with four or five lines of data, a two-line item listing above and below, and all other item listings as one line. This type of interface is best used for situations in which the user needs to get an overview of the data as well as focusing on specifics.

However, the technique can be used less rigorously for advertising, allowing sponsored items to grow and provide more data.

Design

A fisheye ad, illustrated in Figure 6.8, should be based on the highlighting scheme for a list. It needs to feel like a natural extension of the list, and the context of the list must be maintained. For these reasons, the advertisement should be:

- twice as tall as other items in the list
- centered vertically on the middle of the list item, with half of the preceding list item and half of the succeeding list item visible
- the same horizontal width as the highlighting for the underlying list.

The content of the ad can be text-based or graphic-based, but must maintain key parts of the underlying list item.

In general, highlighting can cause a one-line abbreviation of content to expand into two or more lines of text or graphics, either moving content on the screen further down, or appearing over other content. The 'appear over' approach is likely better. Having parts of the screen move outside the control of the user causes confusion (and irritation) due to the unpredictable movement of content.

If the first item in the list has a fisheye ad, move the content below the ad down by one line upon screen display. When the user scrolls off of the first item, revert to normal display. Do not change back to the shifted content even if the user scrolls back up to the first item.

Consider whether the target of an item highlighted by a fisheye should be different from the standard items. The answer varies with the situation.

Applicable Devices and Platforms

It is suitable for the web with scripting and application platforms on scroll-and-select devices.

When Used

A fisheye advertisement must somehow highlight a specific item in a list, usually a search result. This is akin to Yellow Pages advertisements in-line.

To make the advertisement special, no more than approximately 20% of the listed items should have an ad. In a static environment like paper Yellow Pages, this means limiting the number of ads sold. Listings such as movies, particularly with dynamic filters, provide a more interesting environment in which to sell.

Rationale

A fisheye ad provides space-efficient but highly visible advertising to premium listings, opening up streams of revenue for an application without costing user experience.

6.5.3 Banners

Mobile banner ads have come, and gone, come again, left again, and now come again. At some point they will stay in the mobile space, but only if done well.

Design

The Mobile Marketing Association[2] has industry standard best practices for banner ads. Be sure to follow those requirements carefully, and expect the practices to evolve over time. Use the 'less is more' philosophy to avoid banner blindness in which the user's eye simply does not register banner-appearing content.

In addition to following the MMA guidelines, a banner ad system should, where possible:

- make the ad content highly contextual, based both on content and other information discernible about the user
- ensure that the advertising site is as well designed as the native application
- set default focus below the banner ad, so the user does not have to scroll past it
- provide a return pointer on the advertising site, so the user can easily return to your application.

Note that frames are not readily available in the mobile environment.

Applicable Devices and Platforms

It is suitable for all, especially web sites.

---

[2] http://www.mmaglobal.com

7

Graphic and Media Design

Media, including graphics, animation, video, and sound, is perceived and consumed differently on mobile devices. The smaller screen and environmental light conditions affect one set of differences; user behavior and context affect another.

The visual design of a full-sized application, media stream, or graphic image cannot simply be scaled down from its full-sized version. Not only is there a minimum size for text to be useful, there are details visible on a full-sized screen that become invisible on the small screen. A wide-shot picture of a baseball outfielder catching a ball works well on the large screen; if simple resizing is applied then the ball disappears and the outfielder is only 4 pixels wide.

Thus the visual design of any media content has to be rethought to ensure usability and positive affect.[1] The principles of graphic design remain the same for mobile as they do for the desktop, but their context is changed significantly.

7.1 COMPOSITION FOR THE SMALL SCREEN

Designers and artists are accustomed to displaying their work in a frame, whether that is the edge of paper, software and computer frame,

---

[1] Affect: the conscious subjective aspect of an emotion. Positive affects include interest, excitement, joy, pleasure. Neutral affect includes surprise. Here we are using 'affect' to mean the emotional response to an application. Affect is influenced by a variety of factors, including aesthetics and usability.

---

or a physical frame for a painting. The frame is part of the overall design, and helps define what is part and what is not part of the experience. Within the frame, the artist defines positive and negative space to lead the eye.

On a mobile device with a small screen, there is not enough space for negative space to register. It is all positive space. An effective mobile image uses a different composition from a similar image intended for a larger display. A graphic that would look horribly crowded on the PC is clean on the PCD.

Similarly, the use of line and color differ for mobile devices. Consider, for example, the mermaid illustration in Figure 7.1. It is suitable as a story image, as a web site theme, or as a background image. Its color version has subtle shading, pencil-weight lines, and background

Figure 7.1   Art intended for computer display (Elizabeth Leggett)

Figure 7.2 Same art as in Figure 7.1, as displayed on a phone, created by scaling and converting to lower pixel density and reducing to 16-bit color before converting to grayscale. Note that this is likely to be viewed in glare conditions, making the colors disappear further

interest, although only the grayscale is depicted in this book. Were it to be simply scaled down to mobile phone size, as simulated in Figure 7.2, most of the beauty, color, and line would be lost.

7.1.1 Learning from Portrait Miniatures

Creation of content intended for small displays is not new; there is a long practice of painting miniatures in art. Coins can also tell us about design for small displays.

Portrait miniatures, such as those abstracted in Figure 7.3, were the wallet photographs of the time; many of them were smaller than mobile

Figure 7.3 Miniature portraits have many of the same constraints as mobile devices

phone screens. Some were used as lids for tins; others were jewelry. Some had frames, but many did not.

A full-sized portrait of the time would include the full or half length of the body and typically some bit of personalization beyond clothes like a treasured object or a symbol of the subject's status. Miniatures could not contain all this information. Instead, most depicted the bust only; any adornments were worn in clothes or hair.

Full-sized portraits were distant: the viewer is distant from the painting, and the artist adds a more formal distance in the composition. Miniatures were intended to be held, sometimes close to the heart, so the artists painted the subject a bit more intimately.

Miniatures artists used flat backgrounds rather than complex: their goal was to have the face float above the painting. Background complexities would be distracting and likely lost.

Graphic designers can learn from these earlier artists. The user experience for a given graphic or visual design needs to be more intimate when mobile. If creating an image with a foreground subject, then keep the background simple. Crop tightly. Keep details relevant. Play with perspective.

7.1.2 Distinguishing from User-generated Content

American miniature portraits serve a further inspiration: the second generation of such portraits were largely painted by amateur artists. Contemporary professional designers, photographers, and artists may want to take note of characteristics of amateur art.

Camera-phone pictures share much in common with the amateur miniature portraits. The composition is nearly identical: no attention is paid to background, close cropping is normal, and the typical image is head and shoulders. This is in no small part because users are previewing the content on the same screen as the content will be consumed.

With various content-sharing applications becoming increasingly popular, professional artists and designers will find themselves competing with amateur content. Why pay for a special wallpaper or ring tone when there is a tool that lets me convert my favorite photo from Flickr or my favorite song for free?

To go beyond the typical amateur content, a professional should use advanced composition techniques, with careful attention to user behavior. In particular, pay attention to the flow of the eye through the piece, and use creativity in identifying the most important pieces. If attempting to convey worry, for example, consider a picture of wringing hands rather than a face.

For some types of media, the quality of capture and edit technologies will make a noticeable improvement. Quality microphones, cameras, and so forth will go a long way to improving the user experience.

Finally, some types of content delivery mechanisms will be more readily available to professionals. Qualcomm's uiOne, for example, allows for control of wallpapers and other elements of the user interface. This is simultaneously more technically complex than most amateurs can handle and more difficult to get on the device due to carrier protections. Adobe Flash Lite requires an investment in tools and significant expertise, particularly to get small-sized good-quality animations.

7.1.3 Style and Technique

In general, design for the very small screen involves creating, not removing, space. A quick list of principles includes:

- crop extensively
- select for high color contrast
- thicken lines
- limit details, bolder statement in shapes
- forget subtleties
- avoid processor- or memory-intensive designs.

You may want to design in 300 dots per inch (dpi), converting to 72 to test on a device.

Avoid alpha transparencies and gradients rendered by the device. They take up a lot of processor power and are likely to be missed on the small screen. If a gradient or transparency is necessary, create it as part of a bitmap. Avoid requiring the target device to perform unnecessary calculations.

Referring back to our mermaid of Figure 7.4, the figure is already closely cropped. However, the colors would wash out on many mobile devices, so we add new colors to the faces. Knowing that the pencil lines will smudge, we use more color to define shapes. While we keep all the details – indeed, adding some background – we understand that the background details will be lost, that the mermaid will melt into the water. Each of these changes is made without destroying the character of the original illustration.

Figure 7.4 A mobilized mermaid has bolder colors, more skin tone, and a different plot line (colors simulated in grayscale)

Figure 7.5 Cartoon pane

Figure 7.6 Cartoon pane with separable text

Cartoons are a good choice for transferring to mobile content. After all, they are already drawn in small boxes, well cropped, with thick lines and few details. Converting a three-panel comic to a sequence of three panes is both simple and obvious. However, the text embedded in the image will cause problems.

Consider the cartoon pane in Figure 7.5. It is already in a single pane, so it is a good candidate for mobile. However, the text would become unreadable on a mobile device. In general, any text at six points or less will be illegible.

To make this cartoon pane shrink well, the text needs to be separated from the image, and the image recropped to allow space to display the text and keep the image as large as possible. In our example, we've put the text at the top of the panel, but the bottom is also common (see Figure 7.6).

7.1.4 Context of Use

The quality and size of a graphic also depends on its context. A splash screen, home page, or wallpaper, for example, convey both first and repeated impressions. It may be worthwhile to increase the quality, and hence the size, of such graphics. The increased bandwidth or storage load will be counterbalanced by the improvement in experience. Where possible, reuse this graphic or cache the information to reduce its cost.

Supplemental graphics, particularly those seen infrequently or once, usually should be used sparingly. Subtleties should be avoided; information should be communicated in bold strokes. Strong lines and a cartoonish style will do well.

Content graphics, such as the map in a directions program, should be carefully designed to convey the maximum information with the minimum detail. Subtleties can be useful, but only where they add value. Avoid conveying critical information in a subtle fashion, as it may be unnoticed by users in glare conditions. High color contrast is extremely important.

Logo graphics should be small and clean. Consider a mobile version of your logo. The Coca Cola logo, for example, is quite legible if it uses the full screen width. It worsens, becoming legible only due to our long exposure to it, if it is small enough to be embedded in a page. The Coke logo, however, shrinks nicely.

7.2   VIDEO AND ANIMATION

If graphics in general introduce problems, moving graphics causes even more. All the principles discussed in the 'Composition for the Small Screen' section apply.

As video is data-intensive, modern standards do much to compress. Many standards, especially MPEG-4 (including QuickTime), are moving away from the traditional frame-by-frame presentation to instead detect parts of the screen that remain largely static and save those as static images. They then change only the pixels containing the action.

The object model for video transmission, as compared with the older frame model, has many benefits for mobile presentation. Unfortunately, implementation of the model has not been universal. For this reason, and for the sake of simplicity, much of this chapter will use reasoning based on the frame model. Fortunately, if the exact connection in question uses the object model, the user experience will only improve compared to the frame model.

User research has revealed that perceived quality of content is good, as long as the frame rate exceeds five frames per second and the audio rate is at least twelve kilobits per second. Some early services were adequate with one or two frames per second, but only early adopters and technophiles were using the service. Unsurprisingly, content with high action demanded higher frame rates, around ten to fifteen frames per second.

Video on the mobile device is not necessarily about the pictures. Keeping the video quality constant, increasing the quality audio enhances the perceived quality of video.

7.2.1 Content

The mobile television explosion will eventually teach us what users are willing to watch on their mobiles, but preliminary evidence suggests that television, when simultaneously broadcast to mobiles and public broadcast, is generally viewed at home. Users watch their phones while they take a trip to the bathroom or kitchen. Users aren't necessarily watching an entire episode on their phone, but rather short bits.

Content divorced from broadcast television is likely to have to meet a higher standard. Users won't have the context provided by a full television's audio and video capabilities to provide extra richness to the mobile content.

Users are unlikely to want to watch a 30-minute video on their phone, due to connection charges, changing local environment, power consumption, and general attention span. Likely to be more popular is content divided into smaller chunks. The popularity of the one-minute mobile episode (known as a 'mobisode') is still uncertain, but the industry is still immature.

All is not lost for repurposing broadcast television or even movies, but their organization should be changed.

Broadcast television logic, with a long history in the publishing industry, has a hook just before a commercial break to tempt watchers into staying with the show during the commercials. The end of episodes of course have a cliffhanger. Show segments are of course the time between commercials. If this organization were forced into the mobile environment, users would likely feel compelled to experience an entire episode in one sitting.

A mobile-friendly reorganization of television and movie content would respect the fact that the user may take seconds, minutes, or hours between segments. Readers of long novels may be more similar to consumers of long video content while mobile than are television watchers. Many novel readers stop, perhaps for the night, in the middle of a chapter. They know from long experience that the end of the chapter is likely to be a cliffhanger, and make them want to continue reading. The mid-chapter pause also provides significant context for when they return to the book.

One potential organization of a long television episode is thus similar to a chapter of a novel. Chunk the episode into segments of only a few minutes, ending at a low-action state. Put internal cliffhangers in the middle of such segments, and put an advertisement in the middle of the segment.

A side benefit of this organization, for some technologies, is that the next segment can be downloaded while the user is otherwise occupied. The response time when the user requests the next segment can be very short, decreasing barriers to content consumption.

Of course any television or movie repurposing would ideally have the sort of cropping, scaling, and color management described in the previous section on composition.

When composing video content solely for mobile streaming in variable quality environments, consider the role of audio. User perception of image quality is affected by the quality of the audio. More compelling audio makes for better user perception of the video. Use the images to support the audio, not the other way around.

7.2.2   Production and Preprocessing

The better quality you can obtain from your production and preprocessing steps, the better quality the final product will be, even after compressing and reducing for the narrow bandwidth. A professional-grade digital format will avoid quality degradation due to conversion and capture. Like other graphic content, video should be captured at full broadcast size and frame rates, even though it will need to be reduced later.

The most obvious issue with mobile multimedia is the fact that the screens are small. This obviously results in fewer pixels per screen. An appendix discusses mathematics and practices for determining a physical object's resolution on the screen. In general, don't expect much detail. Techniques for handling this include:

- Avoid wide shots when any fine detail is important to the shot. While wide shots are often used as establishment shots, and can be effectively used for that purpose with mobile devices, keep them short. Per the example resolution calculations in the appendix, a scoreboard likely will be unreadable on a shot wide enough to show an entire ball field.

- Shoot more closeups.
- Avoid watermarks. They won't scale down. Text background data, such as crawlers, box scores, etc. will have to take up a much larger percentage of the screen to be effective.
- Better results will be obtained by using a camera and crew specializing in mobile content, rather than merely attempting to reformat broadcasts for mobile distribution. The relative importance of this is determined by type of content and type of audience. Look at the storyboard for your broadcast. Identify from the storyboard the shots that won't work mobile, and the shots that will. This will tell you what is needed in the way of extra resources for a broadcast/mobile combined production. It may be possible to do without any extra cameras, or it might require one or more cameras dedicated to capturing mobile optimized shots. The two storyboards should tell you.

The technology will get better, but these recommendations will largely still apply. Faster, more reliable networks, better compression and smarter pixel scaling algorithms are all in the works. Display devices will have more pixels, memory, and processor power. However, The Carry Principle dictates that the physical screen size has to be small. So, except for projected displays, resolution may no longer be limited by the number of pixels, but by the absolute size of the displayed object on the mobile screen.

7.2.3   Post-production

The type of content you are capturing affects users' perceived quality. For example, a 'talking head' can deliver a good user experience on relatively slow networks with lower frame rates, whereas a sports clip requires a higher visual quality. There are several things you can do to reduce the need for high-fidelity shots, as outlined above. However, if you have audio content it needs to be at least 12 kb/s to support the video. A lower quality audio actually degrades the perceived quality of the video.

Wireless networks deliver speeds that vary from moment to moment, depending on network conditions. A connection could start at near broadband speeds and drop to slow dial-up speeds. Unfortunately, despite the MPEG-4 standard, most networks do not yet allow you to detect network conditions and respond accordingly.

Most networks deliver video content in individual, full-screen refreshes. Achieving the minimally acceptable frame rate is more difficult in this environment. If the media delivery system is using the MPEG-4 capabilities of partial-screen refreshes, you will get higher perceived frame rates.

MPEG-4 compression of motion pictures involves reducing the amount of information that is needed to describe a series of frames by sending full information only once every several frames and using techniques to predict the other frames and describe only the changes from the anchor frame (key frame). Predicted frames follow, and contain much less information than the key frame. The ratio of predicted to key frames should not exceed ten to one.

A variation on this technique is useful for creating animations. The human eye will fill in details that do not actually exist, so the designer can merely hint at those details, using less memory.

Lossless encoding will give slightly better quality than lossy encoding, but between lost frames, low bandwidth, and limited device processor and speaker capabilities, users are unlikely to notice. In exchange lossy encoding gives better compression, which improves the user experience. You may find that lossy encoding allows you to encode at a higher frequency.

As is true with any video process, professional level practices will improve end quality. These include using professional-grade encoding hardware such as SDI or Firewire, and converting from analogue to digital signal at most once. Your standard process will likely cover all these issues.

7.3 SOUND

Customer testing indicates that low-resolution, low-frame-rate images are perceived as better when accompanied by better quality audio. Thus audio quality has a disproportionate effect on the overall user experience. Fortunately, most PCDs are optimized for delivering audio content. Keep in mind that some phones are known for bad audio quality, so not all users will experience the same quality. The Palm Treo line is notorious for this problem.

A further challenge for perception of audio quality is the user's ambient environmental noise is highly variable, both in volume and in frequencies. There is not a lot the content producer can do to overcome ambient noise issues beyond making the content available when the user is ready to hear it.

There exist relatively simply tricks to improve the perceived quality of audio delivered over mobile devices, most related to device and network capabilities. First, many devices do not support stereo playback, and may not support stereo signals. With that in mind, deliver audio in mono unless the device is optimized for music.

Second, use QCELP encoding for speech content. This is the standard that the phones themselves use to transfer speech in voice calls, and is well-optimized for speech in the mobile environment. It is not optimized for music, a fact that can be experienced by listening to music somebody else plays into their mobile phone when they call you. Use AAC for higher-fidelity applications.

Third, user perception of multimedia quality does not directly increase with transmission frequency. A 12 kb/s audio recording delivers a good experience; less is not acceptable and more does not get a significant increase for most applications.

7.3.1 Content

Images support the audio rather than the converse. This makes the audio the primary medium. Make sure that your audience would be satisfied with the production even if it were audio only.

The audio content in a broadcast may not translate well to mobile. Just as you would identify places in the storyboard where you would use alternate camera work for the mobile version, you should do the same for audio content. If, for example, the storyboard calls for 'letting the picture speak for itself', consider adding audio content to support the image.

A large dynamic range can cause a problem. Traditional analogue audio processing techniques to compress dynamic range may result in less efficient digitization of the signal and hence a worse perception of quality. Restricting the range of audio content, like limiting the color palette for graphics, can avoid compression problems.

7.3.2 Post-production

Post-production tools generally help optimize content for mobile, without requiring a lot of specialized knowledge. It is important to note that the more automation you bring into the process, the less control and the more unpredictable the results will be.

Keeping a degree of control over the final product usually will increase the overall quality. Reduce the amount of compression the tools will do by manually compressing your audio as much as possible. Use techniques such as limiting, sample rate conversion, stereo-to-mono conversion, low- and high-pass filtering, and noise reduction to reduce required bandwidth while maintaining the experience you want the user to have.

When using QCELP, use full-rate QCELP audio encoding. The experience is significantly better with a full-rate (13 kb/s) experience over a half-rate (8 kb/s) experience. Users will also perceive your video as being higher quality.

7.4 STREAMING VERSUS DOWNLOADED CONTENT

Streaming audio and video media content can suffer from inconsistent network connectivity, both between the PCD and the carrier, and between the carrier and the content server. The connection speed will vary during a single session depending on network conditions: during peak periods, the connection could slow down to 9600 baud. Managing the balance between network capacity and best possible media quality is partially handled with caching and partially handled with the file formats.

In contrast, downloaded audio and video can suffer from digital rights management challenges, synchronization issues, and PCD storage capacity. It has the benefit of needing to be downloaded only once, and is thus a good solution for content that will be repeatedly experienced.

Designing for streaming content involves all the considerations associated with downloaded content combined with an extra understanding of frame rates in the mobile environment.

Mobile frame rates are low, and streaming mobile frame rates are also variable. If mobile media standards were the same as broadcast television, this would make the video content start and stop as network conditions varied – an experience that happens frequently with streaming video content on many computers.

Fortunately, current encoding tools make good attempts at increasing quality, and this is a case where the small screen works in your favor. Refreshing part of the screen rather than all of it allows QuickTime and some other tools to reduce the amount of data transferred, leaving the available bandwidth for content-relevant screen updates.

You can have some control over video frames per second and audio bandwidth. Network speeds will vary, and the user will receive only a sampling of the frames you send.

Be sure your content will work with only five to seven frames per second, but keep audio fixed at full rate. High-action media may want to check for bandwidth of at least 60 kb/s before attempting to display the content. This speed will allow for ten frames per second.

Streaming frame rates will vary between two and twenty frames per second. Any actions that result in large portions of the screen needing to be refreshed are likely to result in pixelation and smearing. Techniques to handle this issue include:

- Cut between shots instead of panning to follow action, where possible. Cutting between the start and end points of action sequences may lose the action in between, but only results in a single full screen refresh, instead of causing the network to attempt to push eight or ten full screen refreshes down for about a second's worth of action.
- When panning, keep the video subject (vehicle, player, ball, etc.) tightly cropped and centered. The background will likely smear and pixelate but a well-tracked subject may not.
- In sporting events, use replays and slow motion or stop motion to provide a view with compelling levels of detail. Again, keep the video tightly cropped.
- Minimize the effects of rapid motion. Where possible, use shots that move towards or away from the camera.

7.5 MANAGING MEDIA: META DATA

The increased availability of amateur content, and content in general, immediately leads to major issues with content management. This problem is particularly obvious with services like Flickr, which contains user-generated pictures and serves as a sharing service. Old, hierarchical storage and retrieval techniques are no longer adequate, particularly for mobile access.

If making a content access application, be sure to use all available meta data to be able to access it. Users should be able to retrieve data using keywords, tags, similarity measures, and other mechanisms for organization.

Apple's iTunes desktop application teaches us a lot about the future of media management. Each song is primarily accessed by artist and album, but genre, composer, year, and even beats per minute are also stored. In theory, all of these sources are included when the user enters a search term. Further, information like user rating, play count, when last played, and playlist are stored. This combination of automatic and user data provides a rich environment in which media and media collections can be found quickly through a variety of methods.

If making a content generation application, be sure to store data with each image or media clip. As an example, the device can potentially provide the following information about any given picture:

- date and time when the picture was taken
- location of the device when the picture was taken, preferably with address and city
- device owner
- user-added tags (or collections)
- color palette of picture
- the names of other people within the user's social network whose phones were present when the picture was taken
- the user's calendar entry for the time when the picture was taken
- a link to any blog entries the user deems related to the picture.

Additionally, a server component of a picture application can reasonably add tags like:

- names of people in the picture, by recognizing faces within each image and doing similarity measures
- names of any landmarks in the picture
- the person who took picture or created the art, if from a sharing service or messaging.

Not only should this type of data be stored, but – like iTunes – it should be combined in many ways to enable access to media in the way the user is currently thinking.

8

Industry Players

The mobile industry structure affects both what is possible and what is practical in application design. The technology standards might suggest that a feature is possible and indeed standard, but somewhere between the standards body and the device the feature doesn't work as expected.

Business relationships, both those already in place and those to be made, affect what is possible. In the mobile space, the carriers and device manufacturers play varying roles across different markets, with results and processes sometimes surprising to outsiders. This chapter covers some of the history of decisions made by various players as a way to understand what they might do in the future.

For example, when Sony and Ericsson joined forces, they created an organization that could draw from Sony's music brands and resources, making the organization the best placed in the industry to integrate music into their handsets. The fact that, as of early 2006, they have not done much more than integrate Sony's proprietary formats does not mean that they will not develop a truly spectacular music communications device. Such a device will not replace an iPod for aficionados, but could nicely supplement it if the companies decided to use open formats.

The mobile application value chain includes the user, the device in the user's hand, the technologies on the device, the connectivity to the web, services enhancing the connection, applications and web sites, and distribution of applications and web sites. Making all this happen are device manufacturers, carriers, technology platform provides, application and content developers, and content distributors.

Designing the Mobile User Experience   Barbara Ballard
© 2007 John Wiley & Sons, Ltd

The power structure amongst these entities varies across continents and is frequently uncomfortable.

8.1   CARRIERS (OPERATORS)

It is the carrier who builds or leases towers and who decides what communications protocol is being used. These do not greatly affect mobile applications. Operators also decide what network services like voicemail and network-assisted location are used and possibly available to applications. These decisions can significantly affect application design. Finally, operators decide what devices are allowed to connect to their network and sometimes quite a bit about their design. This situation may be weakening, with some carriers becoming more like Internet service providers (ISPs), acting only as 'pipes', with services like voice-over-IP reducing the relevance of traditional voicemail.

Operators also have a close connection to the user in the form of billing. Most operators have enabled adding directly to the monthly bill a variety of third-party services, ranging from media purchases like ring tones, to sodas from a vending machine, to utility bills. This allows small purchases to be aggregated into one payment, which makes users less sensitive to the overall amount and thus more likely to purchase more. Of course, purchasing using the device's account reduces the amount of data entry compared to entering credit card data for each purchase.

The carrier's key concerns are simple. Any device that connects to its network must not jeopardize the network. Any service must not jeopardize the network or the relationship with the carrier's customers. Churn, a term that represents the percentage of a carrier's user base who leave in a given month, needs to shrink (churn rates of 1.5% per month are common). Monthly revenue is key. Beyond these key goals, carriers are likely to have significant inertia when making decisions.

8.1.1   Carriers and Devices

Many carriers subsidize the customers' purchases of new equipment. This makes their customer acquisition costs quite high, and results in a difference in carrier and manufacturer goals. Thus when a user obtains a subsidized handset, the carrier has made a significant investment in that user. This fact, coupled with concerns about churn, led to the requirement that users stay with the carrier for a certain number of months or face a penalty.

The carrier then makes revenue based on monthly use, with greater profits for high voice use, high text or data use, and services. Unique services can make the carrier more 'sticky', and increase the end users' switching costs. Ease of use and overall user experience have a significant effect on how often the user uses the carrier's services. In contrast, device manufacturers simply want repeat sales. Ease of use is less important to them, but manufacturers like Nokia have recognized that predictability of use increases repeat sales. The core Nokia user interface has remained the same for a decade.

This set of conflicting revenue streams, illustrated in Figure 8.1, suggests a much greater motivation on the part of the carrier to have better usability of the device. Indeed, more and more carriers are starting to focus on consistency of user experience across devices – with varying success.

The amount of power the carrier has varies across markets. In markets such as Europe, where devices can be used with multiple carriers by a simple card swap, the device manufacturer has a much closer relationship with the end user. In markets such as Japan, the device can be used only with one carrier, who has a stronger relationship with the end user. The result is that European carriers tend to have less influence over

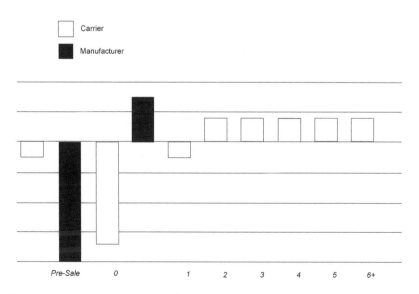

Figure 8.1 Carriers increase profits by sustained use, while manufacturers have to look good on the shelf

device design. Smaller carriers may have no influence beyond deciding whether to allow a given device onto their network.

In Japan, NTT DoCoMo is the leading player in the Japanese mobile industry keiretsu,[1] the 'ketairetsu'. The carriers in Japan are extremely powerful: in addition to providing network service, they specify device design, develop network services, and develop new network standards. Device manufacturers can decide whether to make a device specified by the carrier, but have little influence over its design.

The North American market has both models. GSM carriers tend to get phones available in Europe, and have historically had limited influence over their design. Cingular, for example, did not have a device user experience person at all until 2005. CDMA carriers such as Verizon, Telus Mobility, and Sprint had the opportunity for more influence, especially over Japanese and Korean manufacturers' designs.[2] Verizon was extremely slow to take advantage of this, but Telus Mobility and Sprint have a long history of working closely with manufacturers on device design.

8.1.2   Walled Gardens and Decks

The carrier's 'deck' refers to the set of web pages accessible from the home screen in the browser on the phone. Less commonly, it can also be used to refer to a set of content available from the device's standby screen. It is akin to, but more important than, the shortcuts computer manufacturers install on computer laptops, or the content that Internet service providers host on their home page: less experienced users may not ever discover other content.

Many observers mistakenly call the deck a 'walled garden', but it is not. It is simply a set of content purportedly valuable in the mobile space that will work on the existing phone. Some carriers actively block excursions off the deck, by eliminating the ability to enter a URL from the browser, blocking search engines, and even preventing the launching of URLs from text messages. It is these carriers who have

---

[1] Japanese industry is organized into families of companies across industries who work closely with each other. These 'keiretsu' have dominated Japan for decades, and are one of the major reasons that foreign companies have a hard time entering the market.
[2] European manufacturers like Ericsson (then Sony Ericsson), Nokia, and Siemens would convert existing GSM phones to CDMA. They consider CDMA a niche market, and only provide phones after fully launching the GSM version. Unsurprisingly, they are typically uninterested in changing the design for an end-of-life product.

walled gardens, although to inexperienced users even 'open gardens' may seem walled.

Competition for space on the deck is fierce, and dominated more by money and business goals than by user needs. Regardless, any deck designed as a set of links is limited: each page is limited to around ten or fifteen links, and the user is unlikely to visit beyond three pages deep. That limits the number of sites or applications to around 3000 – an unwieldy and still small number.

As the amount of mobile content available increases, the deck model will have to evolve. Personalization of content – beyond simple bookmarks – will be critical to maintain relevance. Content providers with a niche but desirable brand will stop trying to get on the deck. Mobile search results will become better based on user context, while more sites become mobile friendly. Third-party content aggregators will be able to serve market needs better. All this will increase the users' comfort with acquiring content from sources beyond the carrier.

8.1.3 Mobile Virtual Network Operators

Mobile virtual network operators, better known as MVNOs, essentially purchase network services from traditional carriers. A MVNO could range from the very small, simply rebranding the carrier's services and devices, to the very large, with customized device hardware, separate customer support, separate distribution, and so forth. MVNOs experience reduced barriers to entry into the carrier space, but a full-scale MVNO will still take over a hundred million dollars, US currency.

Companies enter the space either to develop their brand or because they believe they can serve a market segment better than anybody else. The most popular market segment is the youth market, with Virgin Mobile as the success story. Amp'd Mobile is targeting teenage and college boys, and Disney is targeting families with younger children. Apple computer is oft speculated as being a potential MVNO, and there is even a homosexual MVNO, GAYmobile. There are hundreds of MVNOs in existence.

MVNOs fail almost as often as they are launched, but they provide an enormous opportunity for applications to support the target market. Especially promising are social networking applications, given the similarity of people using the brand.

The larger MVNOs specify their own handsets. This is a great opportunity for smaller handset manufacturers like HTC and Sanyo, and makes sense because the MVNO brand is typically larger than the device brand. Customers are attracted to MVNOs based on a promised specialized experience, which is not going to be had as easily with an off-the-shelf Nokia phone. The special handsets can then be used as status symbols, as they give the person a visible membership in a special club.

8.1.4   Network Types

While historically carriers have provided connectivity over cellular (GSM, CDMA, etc.) systems, both new and old operators also provide Wi-Fi hotspots for Internet access. WiMAX will look a lot like third-generation cellular networks, such as CDMA2000 EVDO networks. 2 G connections, 3 G connections, Wi-Fi, WiMAX... all provide connectivity, but with varying speeds, security, capacity, and coverage. Devices are starting to support multiple connection types, which is leading to the need for good algorithms to select which connection to use. Security-conscious users will want the most secure network possible, whereas other users will want the cheapest or perhaps fastest network available.

From an application design perspective, some wireless connections have broadband speeds; others are only dial-up speeds. Thus applications with high throughput needs should detect the network type and either adjust or block the service. Applications would ideally reduce data needs based on cost of connection as well.

8.2   DEVICE MANUFACTURERS

Device manufacturers make phones and other mobile devices. The market leaders are Nokia, Samsung, Motorola, LG, and Sony Ericsson, but they have very different strategies.

Nokia takes a design leadership strategy, with varying physical designs, a standardized user interface, and even careful control over Java environment implementation factors. This has helped them earn the dominant market share. They were the first company to make a user interface that scored well on usability tests. They sometimes fail – the N-Gage gaming console was an example – but not due to poor design. Nokia is also the largest shareholder of the Symbian alliance, and did significant user interface design work to make Symbian a viable operating system.

LG takes more of the attitude of a contract manufacturer, following the leadership of the carriers. Samsung tends to make any device that has a market niche, making largely identical phones with different operating systems.

Motorola and Sony Ericsson attempt leadership roles as well, with varying success. Motorola has made significant investments in user experience for its handsets, but is much better at creating hardware that becomes extremely popular as compared with good software. For example, both the Motorola RAZR and StarTAC were extremely popular and often copied industrial designs that revitalized the company – but users had a hard time finding the web browser in either device. The RAZR was indeed better at finding the browser: an accidental keypress of the up rocker key would launch the browser. Of course, that could incur data charges, so the accident could cost the user.

Unsurprisingly, manufacturers who prefer to take leadership roles tend towards GSM carriers; manufacturers who tend towards a contract manufacturing model tend towards CDMA and Japanese carriers. Smaller GSM operators are faced solely with a 'take it or leave it' attitude from large manufacturers like Nokia and Sony Ericsson.

Selling a new technology to a manufacturer frequently involves selling it to the carriers first. A new browser, text input software, or keypad would have to be approved by the carrier and indeed specified before the manufacturer would do more than create a test mockup to see what the challenges would be.

Manufacturers face the challenge of margins. Carriers negotiate low prices per unit, so unit costs become critical. This problem has driven some manufacturers to avoid operating systems like Symbian, which can cost five dollars per unit in license fees. Samsung attempted to write their own browser to avoid browser licensing fees. Few manufacturers will invest in a clock chip, or indeed any piece of hardware that will not increase the sale price of the device.

8.3 TECHNOLOGY AND PLATFORM PROVIDERS

Platform providers develop the technologies upon which applications run. These providers are in the challenging position of convincing carriers, device manufacturers, and application developers to simultaneously adopt a technology. Sun and Macromedia (now Adobe) have done this by leveraging developers from the desktop environment. Qualcomm has done this by providing a complete ecosystem and a development environment built directly on the chip set. Securing adoption of a platform requires years of effort.

8.3.1 Browsers

Web browser vendors such as Opera, Openwave, AU Systems, Access, and Nokia develop environments to run web technologies. As we learned from the desktop browser wars, the fact that the browsers are built upon standards does not mean that everything works the same. Differences can be due to different technical interpretations of published standards, design bugs, implementation bugs, or the search for differentiation.

Creation of a great browser does not mean instant adoption, as the decision is only occasionally made by end users. For end users to change their mobile browser, they must first use the browser. They really ought to be on a high-end phone as well: while browsers do exist for application environments like Java ME, the overhead provided by the application environment is likely to make a browser less palatable. Next, users must have some dissatisfaction with the current browser and believe that something might be available that is better. Finally, users must be able to find and install the new browser – and keep using it without the convenience of it being the built-in browser. As the number of users who fit all these descriptions is quite small, new browsers such as Opera Mini will have a market share limited to more advanced users.

8.3.2 Application Environments

Operators tend to support any application environment they install on their devices with a complex set of marketing, application acquisition, and developer support. After all, they are looking to increase data usage and to increase switching costs. As a result, only a strategic set of application environments tend to be on a device.

A major complication with application environments is quite the same as browsers: rendering is not the same between providers. In this case, the provider is frequently invisible. Who wrote the KVM for this Kyocera phone? What about this Samsung phone? The answer profoundly affects user interface design, and is hard to discover. One of

Nokia's strengths is that they built their own Java environment, and if an application runs well on one Nokia phone, it will very likely work identically on others.

Other manufacturers rely on third-party vendors to deliver their Java Kilobyte Virtual Machine. They may not specify at all how softkeys are to be rendered or commands allocated. This situation resulted in some Samsung phones mapping all commands to the Menu softkey, including the first item command. The result to the user: the OK softkey is labeled 'OK', and there is also an 'OK' in the menu. The first simply manipulated form elements, such as choice lists. The second actually does what the application developer intended.

When the KVM specifications were created, developers created a reference implementation, which was not intended to go into production. Third-party vendors frequently implement the reference KVM without significant thought to either code or user interface optimization. Obviously the problems experienced with browser variations will exist with Java ME variations.

BREW enjoys a more consistent experience, at least on CDMA devices. The application environment is built directly into the CDMA chip set, and is managed solely by Qualcomm.

8.3.3 Operating Systems

Mobile operating systems of note include PalmOS, Microsoft Windows Mobile (including PocketPC and SmartPhone editions), Symbian, RIM, and Linux. All but RIM are available for licensing on third-party hardware. Thus many manufacturers use the same operating systems. This of course eases porting between devices with the same operating system.

Some operating system providers avoid much of the rendering problems found in browsers and application environments by exerting a lot of control over the implementation. Palm has a finite number of screen dimensions they are willing to support, and all applications at least support a square screen (applications designed for 160×160 generally resize to 320×320 pixels quite well). Microsoft requires most of its manufacturers to adhere to strict standards; manufacturers agree due to Microsoft's strong brand. Even Palm was able to make only a few changes to the Windows OS.

Device manufacturers have three key reasons for avoiding operating systems: high licensing fees, difficulty to put a given operating system onto a specific hardware configuration, and desire for manufacturer differentiation. Symbian addressed the differentiation question well, but did poorly for licensing and porting.

The mobile Linux alliance is likely to address all three issues – but may fail in the ability to port third-party applications across Linux devices. After all, the numerous Linux implementations for computers have user interface and implementation variations between Linux providers. The manufacturers' desire for differentiation may relegate that problem to the application developers.

8.3.4   Hardware and Other Software

Hardware technology providers can have even more trouble in getting their inventions into handsets. A demonstrably easier to use and low-cost keyboard such as Fastap by Digit Wireless was introduced to carriers in 1999; as of 2006 only a handful of devices sported the new keyboard. This is especially interesting since it had low impact on the unit price for the handset, as the predictive text software license could be removed from the unit cost. Device manufacturers deferred to carriers; carriers and device manufacturers were concerned about the perceived complexity of the keyboard.

There are a number of companies providing software to go directly onto the handset, written in the language supported by the chip sets directly. This includes predictive text software, user interface skinning software, and ad serving software. This software has large impacts on the user experience of the phone. The best such software developers work with application environment developers and the manufacturer to integrate their software to give a service available to all functions of the phone. Predictive text in particularly is frequently unavailable to the browser.

These companies frequently get purchased by larger companies. Tegic, one of the original makers of text predictive software for the phone, is now the core of AOL's mobile unit. Trigenix, which provides the carrier the ability to create alternative user interfaces, typically branded, is now part of Qualcomm's BREW group.

8.4   APPLICATION AND CONTENT DEVELOPERS

Application and content developers tend to fall into one of two camps: those extending their content to mobile, and those making mobile their core business. The former tend to be indifferent, and the latter tend to be small. Most content providers have little impact on the operators, manufacturers, or platform providers. Even companies like AOL and Yahoo! are not participating in mobile standards organizations in great numbers.

Note that content can be reused in different applications. The increased use of 'mash-ups' on the web, combining two different web services to provide incremental value over each, can be replicated in the mobile space. Directions and maps are obvious examples of readily reusable content. This content, when integrated into other applications, becomes compelling in a way that it is not by itself.

Application providers have a challenge in getting actual cash flow. Carriers can take over half of the revenues of content sold through their deck, and individual artists can receive pennies for that ring tone or wallpaper. Directly selling content or using independent distributors can reduce sales. This situation is not terribly different from content distribution in other arenas, except the carrier is taking an additional cut.

8.5   CONTENT DISTRIBUTORS

Content distributors, such as Handango and Motricity, started largely by providing storefronts for carriers, device manufacturers, and separate brands, but have grown in services. They have become the best independent experts in device and platform capabilities and rendering differences. Wallpaper and ringer distributors, for example, have to know what formats each carrier allows, what screen size and format each device supports, how to get content to the device, and how to charge for the service.

Mobile search companies are in certain ways distributors. They provide a list of mobile content, can rate the quality of the content or the viability of the content and design for the mobile, and can even establish a billing relationship. There are dozens of mobile search startups, and it's never wise to forget Google.

Independent content distributors will become more popular and relevant as the carriers' role becomes less relevant. The carriers provide two major advantages over independent distributors: payment for content can be readily charged to the phone bill and the content can be placed on the carrier's 'deck'. The content provider can provide a similar service with charging to credit cards, and credit card companies have fewer incidents of incorrect or unsolvable billing errors.

8.6 INDUSTRY ASSOCIATIONS

Industry associations provide the ability for related companies to define industry standard solutions, with an eye towards such goals as interoperability or best practices. Other organizations, such as the CTIA (formerly the Cellular Telephone Industry Association, but now self-described as 'CTIA – The Wireless Association'), do not attempt to generate standards and are thus less important to application providers in terms of direct benefit. Perhaps the most important wireless telecom industry association is the GSM Association. The GSMA develops the GSM family of network standards, but also develops platforms such as SMS and now instant messaging.

Membership in these organizations requires significant financial commitment from a company, and it is accessible only to large companies. Most small companies cannot afford tens of thousands of dollars combined with significant time and travel by highly trained personnel. If the organization is structured to allow small company involvement, they generally allow little more than shared media access and a bit of marketing. Voting is rarely bestowed on small companies.

Particularly important to application providers is the Open Mobile Alliance (OMA), which grew out of the WAP Forum. Original key players were Nokia and Openwave (formerly Unwired Planet and then Phone.com); when WAP was merged with iMode,[3] NTT DoCoMo became a big player. The OMA targets data services and issues, including interoperability, synchronization (via SyncML), presence for applications like instant messaging, mobile commerce, push-to-talk, location, messaging, and of course browser and content services.

Other relevant organizations include the Mobile Marketing Association, which creates standards for delivering advertising to mobiles, and the Mobile Data Association. These organizations, while less influential, are open to smaller companies and listen to application providers. They can provide best practices for specific types of content, and are well worth investigating. Professional organizations like Mobile Monday and Mobile Content can provide excellent learning and networking opportunities for individuals and their companies, but have little or no direct influence over actual practice.

Politics abound in these associations, having sometimes absurd effects on the standards. Notably, Nokia has championed using

---

[3] See the appendices for a history of mobile markup languages, the WAP Forum, the W3C, and the companies listed.

PC Internet technologies largely unchanged, while Openwave has championed making an Internet experience optimized for the mobile device. Their conflicting views have generated lasting impact in the evolving WAP technologies. In particular, WML 2 is a combination of XHTML Basic, some formatting tags, and an 'optional' set of wml extensions to aid navigation. In practice, only Openwave, whose Handheld Device Markup Language (HDML) was the source of the extensions, is the only browser manufacturer to support the extensions.

8.7 GOVERNMENT

Various governing bodies, particularly the United States Federal Communications Commission (FCC), are key industry players. Unsurprisingly, the CTIA as well as individual companies spend significant resources lobbying the government. When Sprint merged with Nextel, one reported reason for maintaining simultaneous headquarters – one near Washington, DC – was to maintain a presence near lawmakers.

Recent issues affect mobile telecommunications:

- Wireless number portability (US) requires carriers to allow the transfer of phone numbers when the user changes carriers. The implication for application design is that the user's phone number is no longer a reliable indicator of where she works or lives.
- Emergency location requires that the phone's location be transmitted to the emergency services response number (911). It was this regulation that forced the US carriers to add location services – not the potential for enhanced applications or services.
- Prodigious patent awards, such as those for 'wireless email', were awarded for reasonably obvious extensions to existing technology – frequently to companies not working on commercialized products. These patents make it difficult for application providers to avoid violating patents. The effect is to inhibit entrance to the market due to likelihood of lawsuit.
- Various indecency regulations focus on adult-themed mobile phone content as being more hazardous to children than normal Internet content. The result is that carriers attempt to restrict content based on opt-in rules; such practices affect all users' access to benign as well as adult content.

Other regulations exist, but have not had a major impact. One such regulation, Section 255 of the Telecommunications Act of 1996, requires telecommunications equipment and services to be accessible to people with vision, hearing, or touch disabilities – where they are 'readily achievable'. Industry players are quite adept at arguing what is and is not readily achievable.

A similar regulation may end up having broader impact, through market forces. Section 508 of the Rehabilitation Amendment Act of 1998 requires the US government to buy telecommunications and computer products, both hardware and software, that are accessible to people with disabilities – if available. While this officially applies only to the federal government, state and local governments frequently adopt federal regulations without modification. Thus this procurement policy can affect 25% of the working population of the United States. Hardware and software that is more accessible, coupled with a demonstration of its relative accessibility to purchasing agents, should enjoy a large government user base. Historically the mobile industry has ignored this policy, but if one major device manufacturer or carrier made a systematic effort at increasing accessibility, the remainder of the industry could not afford to ignore the government market.

9

Research and Design Process

While different organizations have different product development processes, most will have many of the same stages. User experience management adds deliverables, design activities, user research, and user interface design pattern library management.

Figure 9.1 shows a somewhat abstract product development process with user experience design (UXD) deliverables and pattern library information flow. Many organizations' processes are similar.[1]

The chief addition that mobile adds to the process is the need to produce designs for each device class being targeted. The *market analysis* phase, while learning about the devices the target users have and how the users select phones, is the best time to identify common devices.

The *requirements gathering* phase (and not before) is the best time to select both the development platform and the nodes within the device hierarchy to target. Use information about target users, their devices, their training, and their diversity to help determine development strategy. Combine user and device information with project needs, application complexity, and organizational capabilities to decide what set of nodes to target.

A corporate intranet application might not have a large enough user population to justify multiple designs. A least common denominator design might be possible. In some companies, a generic scroll-and-select

---

[1] Organizations using 'agile development' can modify many of the techniques in this chapter for their use. In particular, prototype testing, card sorting, and user interviews are likely to be beneficial.

Designing the Mobile User Experience   Barbara Ballard
© 2007 John Wiley & Sons, Ltd

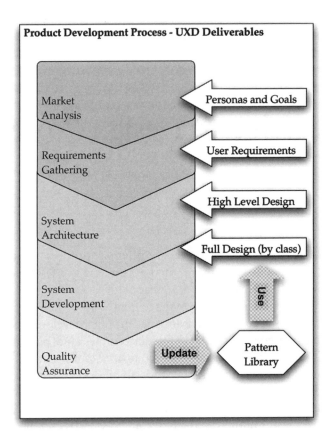

Figure 9.1 A typical software development process with User Experience Design (UXD) deliverables

design might be optimal if few or no employees have PDA devices. A very simple web site is likely to work well with a generic mobile design. On the other hand, a highly interactive application or a frequently used application like a browser or email client will be well-served with a stylus version and different versions for various scroll-and-select user interface paradigms.

A good design program begins with an understanding of user goals and behavior. This type of information is not readily elicited in focus groups. Instead, contextual interviews and ethnographic research are better tools for understanding detailed user behavior and goals.

Typical research activities within the product development process are illustrated in Figure 9.2. This chapter discusses these activities within the context of mobile application design.

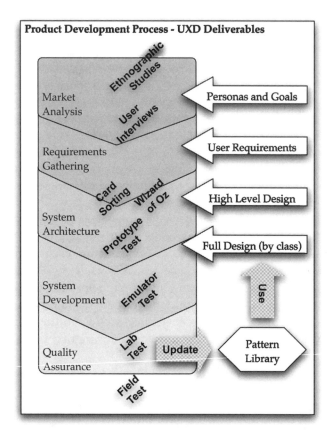

Figure 9.2 UXD research methods, each at the approximate point in time within the product development lifecycle

9.1 MOBILE RESEARCH CHALLENGES

Any usability expert will tell you that the best and easiest method of making your application easy to use is to test it, early and often, and incorporate the results into the design. Various direct research techniques together incorporate the best way to understand user behavior at every stage during development, including storyboards, information design, interim development stages, and beta release. However, the mobile environment introduces several challenges associated with device proliferation, application modality, data collection, and user mobility.

The context of use of mobile applications means that user testing in a nice quiet laboratory or conference room may not extract the most important information regarding an application. Further, the interruption nature of many useful mobile applications (including any that rely on sending SMS messages) renders laboratory testing almost meaningless. Mobile user research has the same challenges as normal user research, with some extra issues associated with application modality, user tasks, and device proliferation.

9.1.1   Device Proliferation

Device proliferation, and associated rendering differences, means that each device displays an application slightly differently. While an application might be usable on one device, it might not be usable on others.

Consider the device hierarchy concept introduced in Chapter 6. In an ideal situation, not only should a device within each targeted node be tested with the design, but a device from each child node as well. Testing with a device from each child node helps identify any places where the design favors one type of device over another, and gives input into improving both the device hierarchy and the incorporated user interface design patterns. It also helps ensure that the application is usable by all users.

Further, each user probably knows how to use their own device and does not necessarily know how to use other devices. Testing an application on a Nokia device by putting a Nokia device into a Samsung user's hands will result in poor usability; putting the same device and same application into a Nokia user's hands can result in excellent usability.

There are two methods for testing on devices. The first is to ask participants to bring their own phones and attempt to use the application on the participants' phones. This method is fraught with risk associated with data plans, phones misconfigured, and so forth. The second method is to select a handful of devices from the children of each node targeted, then recruit participants who use one of those phones. Selecting popular phones will increase the confidence in the results while also making recruiting easier.

Not everything has to be tested on a device. Test certain things, such as under which heading a particular page belongs, on paper or on a simulator. Be cautious when testing labeling this way, since a reasonable label on one device may collide with the operating system labeling on another device.

When doing iterative testing and matching participants to their own device is not feasible, use different devices for each test. Remember that you are looking for major problems, so the differences in devices increase the likelihood of finding problems. You're not trying to make a statistical statement about user behavior. You are instead saying, 'if some users in our test found a problem, then some users will find this problem.'

9.1.2 Multimodal Applications

Mobile applications are quite likely to be multi-modal, and this likelihood will only get higher as network technology both increases bandwidth and allows simultaneous voice and data communications. This makes early testing more of a challenge, because the user cannot simply point at something on the screen and describe what to do with it; there is a simultaneous aural experience.

Voice applications have as many subjective elements as graphical applications, and many issues are difficult if not impossible to test before having an actor record the script. However, it is reasonably easy to test the information design portions of an application before even hiring an actor or prototyping anything.

'Wizard of Oz' testing, described below, can be an extremely effective method of testing a multimodal application. The 'wizard', in addition to changing what the screen does, also provides voice response and can even play recorded sounds.

9.1.3 Field versus Laboratory Testing

Mobile users are likely to be using an application while on the move or with interruptions. The extra tasks of navigating a physical space and interacting with surrounding objects and people presumably affects the user's ability to navigate the application. This fact suggests that applications need to be tested in the field, while users perform real world tasks.

Despite these differences, comparison of laboratory and field usability tests have shown little difference in results (for example, Kaikkonen et al., 2005[2]) in terms of finding design flaws. This may be due to the nature of the application or the type of design problems

---

[2] See Kaikkonen, A., Kekäläinen, A., Cankar, M., Kallio, T., and Kankainen, A., 2005. Usability Testing of Mobile Applications: A Comparison between Laboratory and Field Testing. *Journal of Usability Studies*, 1 (1), 4–17.

in the application, but it does suggest that field testing should be a supplemental rather than primary testing tactic.

See 'Field Usability Testing', later in this chapter, for details on when field testing can provide good results and when to skip the time and expense.

9.2 USER RESEARCH

The object of user research is to get a sophisticated understanding of users' tasks, goals, and context, to make better design decisions throughout the design process. Typical methods include ethnographic research, user interviews, and focus groups.

Ethnography is a research discipline inherited from anthropology. A typical study involves observing potential users' entire context, patterns, practices, and needs associated with some type of concept. A possible mobile ethnography project would be investigating how people interact with and share music, both at home and away, with an eye towards creating a mobile device or application that enhances the experience and fits the needs. Ethnography is particularly good for creating brand new products, as users cannot yet articulate their needs and context for a nonexistent product.

User interviews, as part of the persona creation process, are a relatively inexpensive method of getting some of the same data that a full ethnographic study would accomplish. For most mobile applications, standard interview techniques will continue to work. Some mobile applications, especially those with a social component, will need interviews in a variety of physical contexts and perhaps with groups of friends as well as individuals to capture some of the subtleties of application use.

If creating personae based on user interviews, there may be more mobile personae than there would be for the corresponding desktop application, or the personae may be more complex. There are many different behaviors and goals associated with mobile phones; there are similarly many different behaviors and goals associated with the application. These behaviors need to be combined into either more personae or more complex personae.

Focus groups are common within market research, but do not have good results for informing design decisions. They are even worse for mobile applications, because there is no component of the focus group experience that matches the experience of using a mobile device in different places.

In general, user research is largely the same for mobile applications as it is for desktop applications. The mobile application may need extra information to facilitate good design decisions. Common information includes:

- what type of devices the user has; how often they are changed
- what carrier and type of plan the user has; how often it is changed
- how the user makes decisions on which devices to buy (so future devices can be predicted)
- what applications on the devices are used
- how many people are contacted using the device, and by which methods
- frequency of contacting these other people (to get a measure of how intimate the user is with the device and how strong a social network the user accesses with the device)
- where the user uses applications, whether the user is a social activity
- how the user personalizes the device: stickers, wallpapers, ring tones, applications.

9.3   DESIGN PHASE TESTING

During the design phase, many decisions can be tested with little development cost. Unlike user research, these techniques need some modifications for best practice mobile research.

9.3.1   Card Sorting

Card sorting is an information architecture research tool. It is used to help designers understand how users understand and think about organization and labels for information within the context of a large site. In short, it helps designers put information where the users expect the information to be.

To perform a card sorting test, all of the features of an application, perhaps with a description, are written on separate index cards. A potential user, the test participant, is given the shuffled set of cards and asked to sort them into piles of related terms. There is no constraint on how many piles the participant should generate. The researcher can ask the participant to group the piles into larger groups, and should probably ask for names of the piles.

This is quick research with no technology investment that helps organize the features of an application. It is particularly useful for organizing a web site, but many mobile applications also have organization challenges because less information can be displayed at one time.

When performing this research for a mobile application, be sure to alert the participant that the application will be delivered on a phone. For many applications this may not affect results, but the users' understanding of how information is organized on a mobile phone will affect how they sort the cards for some applications.

It may be worthwhile to compare sorting by users who have different device types, particularly Palm, Windows, and Symbian devices as compared to others. The higher end devices have larger screens which may affect how large the piles of cards become.

Analysis of the test results is the same for mobile as it is for desktop applications, with the addition of investigating any correlation between screen size and card organization.

9.3.2   Wizard of Oz Testing

Wizard of Oz testing is a design concept assessment tool. It allows quick user assessment of the structure of an application without interference from application polish. The architecture and layout is reviewed, not the fonts and colors.[3]

Wizard of Oz testing can be done with paper, pen, and sticky notes, or it can be done with a hand-sketch rapid prototyping tool like DENIM.[4] In either case, application wireframes are drawn and rudimentary interactivity is added with electronic tools or sticky notes. As discussed above, a multi-modal application design also can be prototyped and tested with paper prototypes.

Testing with DENIM is largely the same as other laboratory-based usability testing. Testing with paper prototypes requires two

---

[3] The phrase 'Wizard of Oz' comes from the book and film of the same name. It refers to Dorothy's interaction with the wizard, but she learns that it is just a little man behind the screen. Similarly, the user is interacting with the user interface, but there is no computer system behind the UI - just a little man behind the screen.

[4] Available for free from the University of Washington at http://dub.washington.edu/denim/.

researchers: the normal researcher, and the 'wizard' who acts as the computer. The user points to a control and indicates the intended action, and the wizard adds or removes a sticky note, or even changes out the whole screen (full sheet of paper) in response. Very early stage tests may even find the wizard scribbling new designs in response to unexpected user input.

To modify this technique for the mobile design process, prepare different mockups for different classes of device, as discussed in 'Device Proliferation', above. There should be a different design for each target node in the hierarchy. Ask users what type of device they use, then have them use the corresponding prototype.

Consider using a cardboard phone, with the screen removed, as the frame for each paper prototype. Encourage participants to interact with the buttons, not the screen directly. This will increase the fidelity of the prototype in a crucial area: amount of information displayable at one time.

Wizard of Oz testing *will not* reveal problems with:

- color
- font
- legibility
- text input
- softkey versus screen control allocation.

It will reveal issues with:

- application structure
- item labels
- navigation.

It will reveal some but not all issues associated with screen layout.

9.4 APPLICATION USABILITY TESTING

Once coding of a prototype of the final application has started, higher fidelity usability testing can begin. High-fidelity testing is not necessarily preferred to low-fidelity testing like Wizard of Oz, since high-fidelity testing can focus user responses on surface elements rather than key structure. High- and low-fidelity testing have different roles within the product development process.

Usability testing is done during the design and implementation phase to find issues with the design of the application in time to fix them. This type of testing attempts to answer the question, 'Is there a big problem with the design?'. It can be done with three or four participants in each iteration.

Usability testing is also done when the application is complete. This type of testing attempts to answer questions like 'How good is the application?', 'How does the application compare to our competitors' applications?', and 'What problems should we fix in the next release?' This flavor of usability testing requires substantially more users, sometimes as many as 30.

Usability tests can be performed in a laboratory, with either an emulator or an actual device. Alternately the test can be performed in the field.

9.4.1   Emulator Usability Testing

Emulators and simulators allow coders to view their application on the computer, reducing the number of steps necessary to do unit testing. Emulators have an important but limited role in usability testing, but simulators must be fully tested in their own right before they can be used in product design.

An emulator is not a simulator. An emulator uses the same code as what is run on the target device, but on a computer. A simulator is a separate piece of software that makes a good effort at running the mobile application web site or software. Do not rely on a simulator to provide accurate impressions of your application.

An emulator usability test is run identically to a desktop application usability test. Use the same equipment and lab. Users will interact with the emulator by clicking the emulator buttons with the mouse. You can use any automation software that you normally use with an emulator usability test, since fundamentally the application is a piece of computer software.

As with all usability testing, it is best to match the emulator type with the user's existing phone. The device need not be identical, but should have the same core user interface.

The fact that users cannot pick up the device to use it will affect their comprehension of the application and their perception of the user experience. Despite it being an emulator, it will be an application on a computer pretending to be a phone, yet not a phone. This will absolutely affect test results.

An emulator is likely to reveal issues with labeling, amount of information on each screen, language, core user interface, softkey management, and similar issues. It will not reveal many issues with color, signposting, management of interruption, and especially the feel of game play.

9.4.2 Laboratory Usability Testing

Testing on devices in a usability lab provides a closer experience to the actual application. The use of an actual device makes the experience closer to the actual experience. The user can gesture with the device, hold it closer, move it away, and interact with the user interface as it will be used in actual use. Softkeys will act exactly as they will in the final application.

Data collection can be a challenge. Most usability tests record the user's body language and interaction with the product using two cameras. These videotapes are useful for the product team to understand how users are reacting to their product. However, mobile devices are intended to be held.

The best solution to the problem is to put the device on a sled. The sled has two cameras, one pointing at the device face, and the other pointing to where the user's face is if the user is looking at the device. This allows the device to act very much as if there were no cameras at all, while allowing the researcher to use most of the same tools as she would if it were a computer usability test.

As always, it is best to match the device used in the usability test with the type of device the participant typically uses.

When done well, laboratory usability testing will capture the majority of user experience issues. It generally will not capture issues with navigation of physical and virtual spaces simultaneously, interruptions from other people or incoming messages, social issues, or color in a variety of lighting conditions.

9.4.3 Field Usability Testing

Field usability testing refers to testing outside of the laboratory. While typically this involves researchers following participants and asking them to do tasks, it can equally involve stopping a user at a shopping center or in the hallways at a corporation.

Either field or laboratory tests can involve contrived distractions. Since the device does not belong to the participant, interruptions have to be generated by a corporal person, not a virtual one. For formal, statistically precise usability tests, don't try to introduce distractions into the test. Unless you run dozens of users through the test (incurring a large cost), you'll lose the statistical validity of the test without a discernible benefit.

Field testing, in the form of asking the user to perform real world and application tasks simultaneously, can involve similar problems due to the richness of environment and myriad distractions. Kaikkonen *et al.*[5] compared laboratory and participant-following usability test methods and found that both methods discovered the same design flaws, but the latter method suggested the flaws had higher severity. The field study took twice as much time and money with no particular benefit. This is only one research study, which did not include social components, but it does suggest that this expensive type of testing is not worthwhile.

For informal, problem-identifying tests, however, the mobility of the device helps make the test more realistic than laboratory settings and less expensive than following participants. Instead of asking a potential participant to come into a quiet room and use an application, take the application to a participant wherever that person may be. Many people will welcome an interruption over lunch if it means they get $30 for their troubles.

This flavor of field test can be faster than laboratory testing due to differences in recruiting methods, although videotapes will not be readily created. An entire round of testing can be completed during a single lunch period, and the design can be updated with the results during the afternoon. The chief drawback is the difficulty of matching devices to what the participant has. It may be tempting to simply use the participant's device and compensate for data charges, but many users will not have data services activated.

If you ask a participant to use your application over lunch and the test gets interrupted by a visitor or waiter, that's fine. Take the opportunity to watch what happens when the user resumes the task, and see what difficulties arise. If the application does not cause users difficulties when interrupted, it will be easier to use overall.

---

[5] Ibid.

9.5 MARKET ACCEPTANCE (BETA) TESTING

Usability testing gets information about the ease of use in a controlled environment, at a specific point in time. Many times some sort of field test, user acceptance test, or simple beta test is desired. This is typically run by the marketing group, rather than the user experience group.

Many companies give the product to test participants for a month, then bring participants in to a research facility to answer surveys and participate in focus groups. Mobile applications offer the ability to get significantly more detailed information about actual use patterns and task usability with a little extra investment.

The basic idea is to modify the application server to detect when events of interest occur. When that happens, send the user a SMS with a callback number for a brief VoiceXML survey to elicit usability data. This allows for significant research into real-world task usability, application use context, and many other contextual questions.

As this is not a common technique, this section describes it in detail.

- Determine target questions. The design professional, user research professional, marketing professional, and application server expert should get together to design the study. The design and research professionals should decide the ideal set of tasks to be investigated, and the server expert should help them understand what can be done with the server portion of the application.
- Determine what events on the server indicate that a task has been performed, known as trigger events.
- Generate questions about the task as well as the event, since the user may have triggered the event without performing the presumed task. Keep the voice survey limited to roughly one minute. If the user knows that a long survey will be issued each time a task is performed, it will affect if and how she decides to use the application. Unless you have very motivated users, restrict the number of surveys to one or two per day.
- Modify the application server to notify the test server of each trigger event, including the phone number of the user who performed it. The phone number is known because beta testers are registered and have agreed to receive questionnaires.
- Make survey questions into an easy-to-use VoiceXML application. Include a question about what task the user was attempting, with 'that one' type vocabulary. Likert and semantic differential scales translate well in a VoiceXML questionnaire. Allow free response

only where necessary, as each such question will require extra analysis time.

- Put VoiceXML applications on the test server.
- Recruit participants as for a standard market acceptance test. This is simply an enhanced market acceptance test.
- Send users a SMS when they perform a task that results in a trigger event being recorded on the application server. Ensure that each user is bothered no more than one or two times per day. The survey should only be valid for a few hours, as the accuracy of response will degrade even after one hour.
- Analyze usage trends, including how frequency shifts over time. Does the user increase use? Decrease use? While not every instance will have usability statistics, how does the perceived usability change as the participant gains expertise? Analyze usability trends, modeling the formation of expertise and the learnability of the application features.
- Optionally, perform laboratory usability tests at the end of the beta test period to get further information.

User experience professionals frequently have little input or feedback from market acceptance or beta tests, so cross-department coordination is necessary to implement this type of research. Fortunately the data generated will be quite valuable to marketers as well as designers.

10

Example Application: Traveler Tool

Because performance of a research and design program is beyond the scope of this book, this chapter covers a hypothetical application with a target user well familiar to most readers: the air traveler. This includes both frequent travelers and vacation travelers. As professionals who have participated in the planning and design process for an application know, each section in this chapter represents an abstraction of a document tens of pages long.

This application intends to start where other travel sites stop: its use starts once air, hotel, and car arrangements have been made, and continues until the travel is complete and, if relevant, expense reports have been filed. This suggests the business model of partnering with existing travel sites as an add-on service, for a low cost, with revenue sharing with the travel site once travel is arranged. In this off-deck manner, operator distribution issues are bypassed. The marketing goal for the product is to be 'your mobile travel companion'.

10.1 USER REQUIREMENTS

What users require of such an application depends on who they are, what their goals are, and what their context is. Normally the requirements gathering process includes significant user and market research, but performing such research was beyond the scope and budget of this book.

Designing the Mobile User Experience   Barbara Ballard
© 2007 John Wiley & Sons, Ltd

10.1.1   User Types

The users are air travelers of all sorts. They may have made arrangements using a corporate travel agency, a standalone travel agency, a travel site, or even individual air and hotel companies' sites.

The users have mobile phones and are willing to try the application. That does not mean that they necessarily have a working data plan or messaging plan.

As discussed in Chapter 9, personas are a good method for representing user goals and context. In this application, interviewing frequent and occasional travelers, both business and leisure, should provide a good understanding of all users. This does not imply that four personas would result from the research. More likely is two personas.

Beyond frequency and purpose of travel, perhaps the most important distinction for travel behavior is those who heavily plan versus those who do not. Let us thus *hypothesize* three personas for our application:

- *Justine* jumps on an airplane a few times per month for business travel, and only occasionally has time to do any tourist-like activities. She travels so often that she rarely has time to do much advanced planning or research; indeed, she simply reserves the air, hotel, and car. This has gotten her into trouble in places that really do not lend themselves to auto travel.
- *Juan* is planning a vacation in India, traveling the Palace on Wheels for his family. He has never before been to India and does not get a chance to travel often. He is excitedly researching not only the history of the Palace on Wheels, but the opportunities at each stop and potential experiences both before and after the voyage. His family, especially his teenage daughter, hope to keep in touch using text messaging while they travel.
- *Georges* (secondary)[1] lives in Chicago and visits his girlfriend in São Paulo bimonthly. He is well familiar with São Paulo, although he has never lived there.

---

[1] Georges is unlikely to use the application due to the frequency with which he travels to a single place. He is kept in the persona set, as a secondary persona, until we further evaluate his needs and see whether they conflict with Juan's and Justine's needs.

10.1.2 User Goals

Justine, Juan, and Georges have some clear goals.

- All would like to use a mobile phone, inexpensively, at their destination. They would like their phone number to continue working so they can receive calls.
- All would like to minimize waiting time at airports and minimize security hassles, but would like to avoid being late for flights or meetings. Justine is more concerned about waiting time, whereas Juan is more concerned about making his flight.
- All would like to negotiate the new city without getting lost or scammed.
- Juan would like to ensure that his vacation happens without unnecessary difficulties, so his family can focus on the beauty and culture of their destination.
- Justine would like to have as few hassles as possible, and get to her various appointments on time.

10.1.3 Devices

All of our personas have PCDs with text messaging enabled. Perhaps Juan's daughter has a Sidekick. Justine may have a high-end device like a Symbian, BlackBerry, or Palm device. Juan and Georges, and perhaps Justine as well, are likely to have 'mass market' devices with a browser, text messaging, camera, and Java environment. They do not choose their devices based on which API or browser the device has.

If she is doing international travel, Justine probably has a device that supports international roaming on a GSM system.

Most specifically, only a few of the devices will have the Java ME PDA API, which allows Java applications to use the calendar data on the PCD.

10.1.4 Key User Needs

There are six stages of use of the potential product: planning, last-minute planning and packing, actual travel from house to ground transportation, at-location travel, packing and getting to the airport to return home, and post-travel reconciliation. In each, user needs differ. While this product may not serve each of the needs identified, it should nevertheless serve most of them to earn the title 'mobile travel companion'.

When planning the trip, once hotel, air, and car arrangements have been made, users need:

- to get pocket cash and travelers' cheques in the destination currency while still at home (Juan)
- to know whether their mobile will work at their destination (Justine, Juan)
- to know the cost of operating their mobile at their destination (note that a US phone operating in the UK can cost USD3 per minute, with per minute billing) (Juan)
- to know whether mobile data services will work at their destination, and the cost (Justine, Juan)
- to make any necessary plans to get mobile voice or web services (all)
- to understand any cultural differences (appropriate dress, business card etiquette, and haggling practices, for example) (Juan)

When packing and doing other last-minute planning, users need:

- information on any updated airport security arrangements, particularly carry-on restrictions (all)
- current traffic conditions on the likely routes to the airport (all)
- current security delays (all)
- flight status (all)
- when to leave the current location, based on all of the above, to make the flight (all)
- to know what records need to be kept for any relevant tax refund for non-citizens (Juan, Justine)

When traveling to the main destination, users need:

- to get pocket cash in the destination currency (Justine)
- to acquire any necessary supplemental telecommunications equipment (Juan)
- gate information, name brand restaurants, stores, and bars in the airport (Justine, all)

When at the main destination, users need:

- to get from the airport to the next destination, whether that is attraction, hotel, or meeting, without getting lost (all)
- to be able to select ground transportation, such as taxi or shuttle, that is reliable and appropriately priced (Justine, Juan)

- to pick up a rental car with as little delay as possible (Justine)
- to be able to take pictures, either to supplement a separate camera or as the primary camera (Juan)
- to be able to post pictures to a web log or Flickr (Juan, Georges, sometimes Justine)
- local information regarding restaurants (Justine)
- local information regarding directions and routing, either using public transit or private car (Juan, Justine)
- local information regarding when to leave to get to the next destination on on time (Juan, Justine)
- local information regarding last-minute entertainment (Justine)
- where to get cash in the local currency, at low prices and favorable exchange rates (all)
- the nearest location to get cash, at any price (Justine)
- where to get coffee (all)
- where to get wi-fi (Justine, Georges)
- information about points of interest, with related recommendations or warnings (Juan)
- to make voice calls at low prices (Juan)
- to know when voicemail on the normal phone is received, and be able to respond if necessary (Juan)
- to make voice calls at any price (Justine)
- to receive calls at the normal phone number (Justine, Georges)
- to continue to receive instant messaging (Juan's daughter)
- to be able to use the application (all)
- to be able to handle emergency needs, such as acquiring clothes if luggage is lost, finding a doctor, or contacting a repair service for a broken vehicle (all)
- to avoid high data roaming charges (Juan, Georges).

When packing for the return trip and traveling, users need:

- information on any updated airport security arrangements, particularly carry-on restrictions (all)
- knowledge of typical time needed to return a rental car and get into the airport (Justine)
- location of a gas station for less expensive refueling (Justine)

- best driving route to airport and current traffic conditions on it (Justine)
- best public transportation route to airport and expected duration (Georges and sometimes Juan or Justine)
- how to get a taxi (Juan, Justine)
- current security delays (all)
- flight status (all)
- when to leave the current location, based on all of the above, to make the flight (all)

When reconciling their travels, users need:

- to post any trip diary, pictures, or summary onto web log (Juan)
- to fill out expense reports (Justine)
- to track lost luggage (all)

Most marketing departments would blanch at the above lists, but most of the data is available.

Upon review of the user needs, Georges' needs are always either Justine's or Juan's needs. Georges' primary need that would be unmet by a product designed for Justine and Juan is an understanding of public transportation routes and time estimates. We are thus largely safe ignoring Georges.

10.2 PRODUCT REQUIREMENTS

Some of the above key user needs lend themselves to push technologies, others lend themselves to pull. Some lend themselves to desktop web presentation, others to mobile presentation.

These product requirements make the assumption that the primary distribution method will be by an add-on purchase using travel planning web sites like Orbitz.com. The price includes all SMS charges. It may not be affordable for users without a data plan.

This model also addresses rights management issues. When the user purchases the application and enters a mobile number, a WAP Push message should be sent to the mobile. The included URL encodes key parts of the user's itinerary as part of the download, eliminating significant user data entry and errors. The application is only good for the trip for which it was purchased. Future versions, targeted at Justine, would enable a subscription model for all of Justine's travels made through a specific web site.

10.2.1   Features

The product has four clusters of features mapping to the key user needs: travel logistics management, travel alert push, coverage planner, and customized travel portal.

Coverage Planner

The coverage planner module integrates with services like Orbitz and Travelocity, and is also available for travel agents. It is a web-based application, available from both mobile and desktop. It is a free service and helps sell the full application.

The coverage planner extracts the information from the travel site or, optionally, an email or other text with an itinerary. It asks about the user's current carrier, phone model, and services used.

* Based on the information provided, the coverage planner determines whether the user's current phone and plan will be usable at the destination and whether any extra charges are likely to apply.
* If relevant, the coverage planner determines alternatives, including phone rental, prepaid services, SIM swap, VoIP, and simple roaming, for use during the trip. The options are presented with the assorted relevant costs.
* The user may rent a phone, possibly using a third-party company, from the coverage planner

If the user has data services, the coverage planner will offer the application, with the note that it will enable the user to keep up with instant messaging, email, voicemail, as well as providing critical travel support. If the phone is being rented, the setup process for the rental should include preloading the travel application, with all the user's data.

Travel Logistics

The travel logistics application is a mobile application with a supplemental web configuration service. During the setup phase on the web site (just before and after the purchase is made), the application collects all the information gathered by the coverage finder, and:

- any phone numbers, instant messaging accounts, and email addresses that should be monitored while traveling, including passwords (temporary passwords work)
- any known events planned for the trip, including addresses and phone numbers if relevant.

Using this information, a customized version of the travel application gets downloaded to the device.[2]

Both the downloaded application and the web site during setup should provide:

- predicted wait at security
- carry-on restrictions
- necessary documents for travel to the destination
- any relevant document retention recommendations, such as that for VAT refund
- exchange rates and best methods of acquiring the currency.

The downloaded application also provides:

- updated security delays
- updated carry-on restrictions
- ground travel options at the destination airport, with expected prices
- taxi options from the airport
- directions and map to the next destination on the itinerary, including necessary departure time based on method of transport and traffic conditions
- public transportation maps, schedules, and routing assistance
- gas station locations, including prices if available, both nearby and near the airport
- airport details, including which airlines are at which terminals, time required to return cars, and a good directory of services for the airport including all the coffee shops and bars
- access to the travel portal.

Travel Alerts

While the user is traveling, push to the PCD being carried (rental, temporary, or normal):

---

[2] The downloaded application, with updated user input, will also be pre-loaded on any rental device.

- all State Department (Foreign Office, diplomatic corps) alerts from the user's home country for the country being visited
- all alerts for travel through airports the user will be using on the return flight
- all updates to security procedures for travel originating from the relevant airport
- when the user needs to leave for the next destination included in the itinerary, based on current traffic conditions, to be on time
- if subscribed, voicemail notifications
- within 24 hours of travel, flight status updates

Security condition updates, flight status, and departure timing alerts are also pushed to the user's phone during the last-minute planning stage.

Travel Portal

The travel portal is intended to provide access to all the information the user may need during the travels. It provides information only for the regions to which the user is traveling, and only events available while the user is present.

The portal resides on the device as an application with as much data as possible resident without accessing the Internet. Portal information includes:

- emergency services: hospital, auto repair, clothing replacements, and consulate offices
- restaurant, bank, and ATM locations, with directions and relevant fees
- wi-fi locations, including both free and pay
- coffee locations
- instant messaging, SMS, and voicemail, sent to the user's normal contact data, available to be played and replied to from the application(notethat this should collect messages for all family members)
- a voice-calling widget that calculates the least expensive routing to contact a given phone number, including voice-over-IP calls
- a widget to collect journal entries and pictures, with the ability to post to a blog account
- an expense journal, used both for corporate expense reports and tax issues and suggestions for local tipping
- points of interest, both nearby and within easy travel, including directions and map

- entertainment and cultural events, both nearby and within easy travel, including directions and a map
- access to any geotags, local information, or history about the larger area or the current location
- cultural primer, including whether and how to haggle, and other customs
- safety tips, customized for the city being visited
- access to the travel logistics application.

While meeting all of the above user needs would be overwhelming for a first launch of a product, most of the information is available from a variety of third parties and can thus be included at least temporarily while full functionality is developed.

10.2.2 Technologies and Platform

For alerts, some form of SMS should be used. This may be pure SMS, WAP Push, or a combination of the two. Most professionally run SMS gateways can handle either.

For the configuration and planning process, desktop (and mobile) web access should be enabled. As this process will frequently recommend renting equipment for travel use, the web section can even be free. This portion can serve the triple purpose of marketing equipment rental services, marketing the application itself, and serving user needs. The business model and user experience again interact.

The actual application, including logistics management and customized travel portal, should be some sort of downloaded application with web access. Ideally the portal could be temporarily placed as the standby screen of the device, but the need for broad device access is more important. Since BREW devices can run a KVM and hence Java ME applications, there is no reason to choose BREW with its smaller number of covered devices.

The choice is between Flash Lite and Java ME. Java has broader coverage; there are carriers that do not support Flash Lite. Java also provides broader access to device capabilities such as location, which can enhance the user experience when available. Java ME is likely the best technology choice for the application, although Flash will also be acceptable.

10.2.3 Device Classes

Users will have little training on this application, but will have some degree of sophistication regarding mobile phones. At a minimum, target three device classes:

- stylus-driven devices, including Palm, most Linux, Symbian UIQ, Windows Mobile Smartphone Edition
- Nokia-style softkey devices, including all Nokia devices with softkey as well as a number of devices that emulate Nokia
- other softkey devices.

With regard to the latter, users of these devices, if they have used applications, are accustomed to applications that do not fully match the native device user interface paradigm. A Nokia-style UI will take too much power away from them, but minor mismatches on softkey labels will not seriously detract from the user experience, particularly if Java ME's abstract commands are used rather than direct softkey labeling.

The softkey devices are more common, but users of the stylus devices are more likely to download applications and use data services.

10.2.4 Development Strategy

Only a small portion of the functionality in the application will need a custom screen to achieve its information organization goals. Java ME MIDP 2 high-level screens, such as Form, would allow for quick development. Unfortunately, the branding needs of almost any application prohibit use of the relatively ugly high-level screens.

To get the color, font, and layout control of the low-level Canvas while maintaining the speed of deployment of the high-level Form, use a system like J2ME Polish. This system allows CSS styling of high-level components, as well as easy management of device clusters. It currently does not allow for correct allocation of abstract commands on stylus devices and non-Nokia softkey devices, but this application does not target different user interfaces for non-Nokia softkeys.[3]

---

[3] J2ME Polish 2.0 should fix these shortcomings, allowing for native placement of commands. As this is an open source project in beta in the second half of 2006, I would advise clients to continue to use the older 1.3 or 1.4 until 2.0 has been well tested.

The cost of having a pretty and rapidly developed application will be manual control of softkeys for non-Nokia devices.

Where possible, the MIDP 2 CustomItem will be used before a Canvas screen is used. This should help code and design consistency as well as development speed.

10.3   HIGH-LEVEL DESIGN CONCEPTS

During the high-level design phase, core organization and user interface paradigm issues are resolved. Sample screen shots can be created, but hundreds of detailed decisions are left for the next phase. This section is shorter still than a normal high-level design document: no research has been done, many details that would normally be specified are not, and few wireframes have been created. Thus this is about high-level design concepts, not high level design.

For the sake of simplicity in the remainder of this chapter, we will ignore the complexities associated with dual handsets.[4] Alerts during travel itself should go to the normal phone until that phone will not work, as calculated by when the airplane that takes the user to the new environment takes off – but not while the user is still traveling between airports in the home country.

This application has a large amount of information, with varying degrees of applicability based on time. If the user's return flight is not for three more days, flight status is not relevant. If the user just got money, she does not need to know where the nearest ATM is. The quantity of information combined with the amount of control of the information has the following design implications:

- The need for just-in-time information is balanced by the need for access to all the features at all times, in a predictable manner. This combination suggests the need for a well-designed hierarchy.
- Available upon opening the application must be a small amount of contextually relevant information, which should be updated in the background when the application is not running.
- Some card sorting research should help design the hierarchy.

---

[4] The application may have to run on two different mobile devices if one mobile device has to be rented to work in the travel destination. If at all possible, the rented device should have the same user interface paradigm and should have the application and user's data pre-loaded, perhaps protected by a password before it is unlocked for the trip.

- The application should have a page-based design for the majority of its screens, with content allowed to scroll up and down (only), beyond the confines of the screen. This allows a broad array of information to be displayed with little need for editing to fit on a single screen.

10.3.1 Task List

The logistics phase, especially before and during travel, is heavily task- and time-focused. One concept – that needs to be tested with users – is to organize the logistics portion of the application as a task list.

Many tasks would be pre-entered, such as 'Pack for trip', 'Depart for airport', 'Park car', and 'Get to gate', and the user can add extra tasks with or without a due date and time. The user can mark a task complete, ignore it for this trip only, or delete the task from all trips. The combination of task status as reported by the user and outside information such as flight status provides the application significant context. If, for example, the user has not yet marked 'get to gate' complete and the flight is currently boarding, the task turns red and has other information on the main screen that indicates the urgent status.

Each task displays any alert data, such as security advisories or flight status, directly on the main screen of the task list. Further details are displayed when the user opens the task. Details can vary by task, and include maps, directions, and critical information, so that the task details contain all information that the user needs to accomplish the task.

Some tasks may have a user-editable task list in the details. In this fashion, the user can add a packing list directly within the travel application. Other tasks, such as 'park car', might have a field that encourages the user to jot down details on where exactly to find the automobile later.

Available through the menu, either the Options menu on a Nokia device, or a softkey labeled 'Menu' on a standard softkey device, includes Quit, Options, and Travel Info. Travel info may also be its own task – 'Visit destination' – but its details would have a button to take the user to the 'At Destination' main screen.

10.3.2 Communications Center

The communications tool collects messages from various sources as configured by the user, including voicemail from the normal mobile phone, home voicemail, email, instant messaging pings, and SMS from any desired phones if possible. This information would be stored on the application's servers if a live connection to the messaging service is not possible. A chief benefit of accessing the messages through the tool is the ability to use VoIP rather than international phone calls and perhaps roaming data rates.

The communications center is a travel-focused unified messaging application within the larger travel tool. It uses a simple *list-based design pattern*, with commands available for each message, copying other unified messaging user interfaces.

Some filtering may be desirable, such as ignoring all work emails while on vacation, but should be kept simple. Messages can be deleted, forwarded, or replied to; all content is also returned to the home server so the history of messages will not be lost when the user returns home. There is not a major contacts list, but most messages will not need it. Frequent users should have access to any stored online address book, especially Gmail, Yahoo!, and Plaxo.

If more than one family member has been set up, access to each set of messages should require a passcode to maintain privacy.

10.3.3   Maps, Directions, and Transportation

The transit tool helps the user find public transportation, taxi stands or phone numbers, public transit maps and directions, and pedestrian and driving directions. Where possible, information about alternatives such as airport shuttles should be used.

Users will rarely know the exact address of their destination; they will rarely know their current address but it is discoverable by talking to people nearby. We therefore expect destinations primarily to come from searches, itinerary, SMS messages, or (if present) addresses of contacts.

The screen should have the following elements:

- Title – 'Transportation'
- 'Currently at <display address>' – If the user's location is unknown, instead display a button labeled 'Find me', taking the user to a *location selection pattern* screen, ideally with the addition of picture input. Other 'favorite' locations include the airport, the hotel, and any known stops on the itinerary; a 'Search . . . ' item takes the user to the local information tool to find a specific location. If the user has, within the past hour or two, indicated location by a search by

way of picture of a landmark or similar, use that implied location as a default.

- 'Transit method' – and a set of radio buttons including 'car', 'taxi', 'walking', 'public transit', and 'other'.
- (optional) 'Destination' – and a drop-down list of different fields based on radio button selection.
- A 'Map' button is displayed if transit method is car, walking, public transit, or other. On the map is the route to the destination, if any, optimized for the method selected. The public transit map has relevant stops or stations that can get the user to the destination, or all stops and stations if no destination is indicated.

A set of phone number links to be able to summon the service is displayed for taxi and other. If taxi stops are nearby, a 'Map' button is also displayed.

A 'Directions' button is displayed for all methods except taxi, with directions optimized for the method selected.

Maps and directions should, as a base design, copy the designs of Google, Mapquest, and Yahoo!, who have significant experience designing directions and maps for mobiles. This is especially important given that the data source for maps and directions may very well come from one of these companies.

On any displayed map, commands to add ATM, Wi-Fi, coffee shop, and restaurant locations are available.

10.3.4 Journaling

The journal tool records pictures, expenses, and notes for the trip. This should be a simple screen, focused on recording an entry. A drop-down list categorizes the entry into expense, blog, notes, and any user-defined categories. Subsequent fields allow user entry; a softkey saves the data and posts if relevant and configured. Another command shows previous journal entries on a separate screen.

The other fields vary based on category. An expense entry requires type of expense, vendor, and amount, with the confirmation alerting the user about potential international tax implications. Most other entry types use a subject, body, and ability to attach a picture (and perhaps a sound recording).

At the end of the trip, the journal is sent to the user's email, in a standard format like comma separated values. Each category may have its own file.

10.3.5   Local Information

The purpose of this tool is to get information either about where the user is, or about where the user would like to go. Since text entry is relatively difficult and there is significant information embedded in the world around the user, the camera can be an excellent input mechanism.

Ideally, this tool would have:

- Title – 'Find Info About...'
- An item labeled 'Landmark, building, text, object, or code' – with a button labeled 'Get picture'. The camera viewfinder is activated, and the captured picture is sent to an image recognition server. Text is captured, translated if necessary, and searched. A landmark or building serves as a method to get precise location as well as return information about the landmark or location.
- A text entry field with no label, but default text 'business or topic' and a button labeled 'Search'. If appropriate, search terms are translated into local language as well as possible and added to the search.
- A list of links for categories of information, like 'Wi-Fi', 'Food', 'Culture tips', and so forth. This information must be categorized and labeled based on results of the card sort research and should include as many items from the user needs section as possible and relevant.
- A category called 'Money' – with ATM locations. The current exchange rate is displayed next to the category name.
- A category called 'Emergency Services' – which includes consulate address and phone number, nearby hospitals, sources for non-emergency medical care, pharmacies if appropriate, police, and possibly nontraditional services like clothing if users categorize them here during research.

This leaves the main screen of the tool with a picture button, a text entry field, and a well-designed hierarchy based on user research. This can be designed as a simple form. The results should use a *returned results UI pattern*.

In all cases, the search fetches results based on best information about the user's location. At a minimum this includes the presumed city based on the itinerary, at best it includes GPS data. If the user has entered anything into the text entry field before selecting a category, results are sorted by degree of matching the search terms.

Search results will be of many types. If a word or phrase was entered or photographed, it could be a street sign, the name of a landmark, the name of a hotel, the name of a business, a cultural practice, a food type, or many other things. Another key area of user research is to understand what types of things are frequently searched, and ensure that the search results are optimized based on typical needs. A search for 'Italian', for example, should return Italian restaurants near the top.

Certain result types should have actions associated with them. Locations, whether they are businesses or restaurants or ATM machines, should have the ability to get walking, driving, or public transit directions to get to them (ideally with transit time for each). Ideally the name or address of the destination should be available as an audio clip in the local language, for communicating with taxi drivers.

10.3.6 Main Screen

User needs with this application vary significantly based on whether the user is at the destination or not. Before and during travel, the tool focuses on security, logistics, and travel support. At the destination, the application focuses on being a travel portal, with access to logistics.

This information can be gleaned by a combination of time, itinerary, and flight status, although it will not be possible to distinguish packing for the return home from a last night out. Fortunately, most of the information a user might need for packing for the return flight will be delivered by SMS.

Three basic approaches to serving user needs are to have two different applications, to have two versions of the main screen, or to have a main screen that manages to serve both sets of needs simultaneously. If the latter can be done well, it will provide the benefit of predictability for the user and reliability in case the user context is misjudged. We thus turn our creativity towards a unified interface.

Fundamentally, the application consists of five major components:

- the task list tool
- the communications tool
- the transit tool
- the journal tool
- travel tips and local info.

The four tools are each small applications in their own right, whether they are available separately or not. The travel tips and local information are essentially a local search enhanced by knowledge of user context.

One frequently used method for displaying several tools and information sources on web and some other desktop applications is the 'portal'. This is a web page that contains summary level information from a tool, and access to more detailed information. The summary information is in a 'portlet', or a small portion of the screen displaying content from a particular tool. Some portlets are interactive; others are not.

The benefit of a portal is that it provides dashboard-like overviews of the user's information. It does not readily lend itself to mobile, as the mobile is large enough to display just one portlet before scrolling is necessary.

One possibility is to combine the concept of a portal with the concept of a fisheye design, as described for advertising in Chapter 6. Each portlet is the full screen width, and has a design when minimized and a 'full' design. Support for this dual-size concept is built into Java ME's Form classes, but a CustomItem will be necessary to support it.

Figures 10.1 and 10.2 show wireframes for a portal designed in this fashion. In Figure 10.1, a scroll and select device displays the portal with numbered access for each portlet. Items within a portlet can be individually scrolled to and acted upon using up and down. If the user

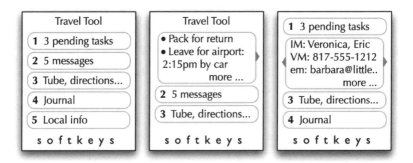

Figure 10.1 Fisheye portal design for a scroll and select device. The left pane shows no item highlighted, and all portlets minimized. The center pane shows the result of scrolling down one click, highlighting the first portlet and expanding it. The right pane shows the result of scrolling down a second click, with the first portlet minimized again and the second portlet expanded.

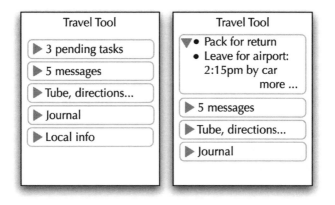

Figure 10.2 Fisheye portal design for a stylus device. Instead of scrolling or numbers to open portlets, reveal triangles are used

scrolls down after the last item, the portlet is exited, minimized, and the next portlet is entered and resized. Left and right scrolling quickly navigates between portlets without having to scroll through individual items; a small graphic to either side of the portlet alerts users to this fact.

Figure 10.2 shows variations for a stylus device. While the scrolling behavior should still work, numbered access is not relevant. Instead, a disclose triangle, common to desktop design, reveals and hides portlet content. The user can expand a portlet by clicking anywhere in it.

In both cases, only one portlet can be expanded at a time. If the user clicks 'more', the tool page is displayed. The expanded portlet should be smaller than the screen height, but should be centered on the screen unless at the top of the page.

10.3.7 Softkey Strategy

For this application, most of the screens will be the same across device classes, but softkeys and buttons will be handled differently for each class. Each screen, and sometimes different items on the screen, has a set of commands that needs to be allocated to buttons, softkeys, and menus.

These commands could theoretically be Java ME's abstract commands, but that could force coding of Canvas screens, as discussed previously in 'Development Strategy'. Instead, commands will be allocated as follows:

- Nokia-style softkeys: put all commands in the Options menu, with the most common command first. Commands should be sorted by decreasing frequency except where there is a strong natural grouping.
- Non-Nokia-style softkeys[5]: put the primary command, usually the most common, on the left softkey. If there is only one other command, put it on the right softkey. Otherwise, make the right softkey be labeled 'Menu', and place the remaining commands within the menu. This menu will be replaced by an identical looking menu with text input controls when text input is active. Backwards navigation will be on the Back button when available; otherwise it will be in the right softkey menu.
- Stylus devices: use abstract commands, using exactly one ITEM command for each interactive item.
- On screens with only two navigation commands, like dialog boxes, the softkeys will be overridden to display the two navigation commands. The primary (usually left) softkey will be the forward command; the secondary softkey will be the backward command.

10.4   DETAILED DESIGN PLAN

Once the high-level design is complete, the design is not finished but some of the decisions that have been made can help the development organization start its own architecture and planning. It is at the detailed design stage that the three target device classes start needing a lot of separate attention, especially the more fully featured stylus-driven devices.

10.4.1   Process

The high-level design should be reviewed, edited, and approved by the design team and the marketing, business development, and development teams.

---

[5] Again, J2ME Polish 2.0 should provide native placement of commands, making the softkey strategy be 'use abstract commands' for all types of devices. This will not negate the need to focus on softkey strategy, as some devices do not render them well, and the commands themselves need to be designed.

The design team should focus on each device class in turn rather than all three simultaneously. A simultaneous focus may result in mental shortcuts being taken, and a design shared between classes simply because it is 'good enough'. Focusing on one class at a time is more likely to result in an optimal design for that device class. What can usefully be reused between device classes should be, and each design will likely influence the others. Do not break the design team into three subteams working independently, or the three designs will diverge, increasing costs.

10.4.2 Tasks

Among the myriad questions remaining to be answered in this stage, the design team must determine:

- how large each portlet is allowed to be based on screen size, amount of content, and user context
- types of data displayed for each type of search result
- precise specifications for what is displayed in each portlet in each context
- how to deal with different data sources
- exact softkey and command design for each screen for each device class
- how to get user data back into normal tools after the trip.

10.4.3 Data Sources

A major challenge with this application is the varying data depending on which city the user is in. Several organizations provide local information. The State department, international business travel organizations, and some other organizations provide information on culture. A company like Citysearch provides guides for many US and Australian cities, AOL provides information for US and Canadian cities, India Catalog.com provides information for Indian cities, and so forth. Many if not most cities have one or more locally run sites which may have more complete information than the larger aggregators.

Within the detailed design phase, integration of each of these disparate sources will have to be designed. Certainly working with a set of large operators, each with several cities covered, will speed the process along, but adding locally generated content may be desirable. Alternatively, simply link to the locally generated site after verifying that the site will work in mobile browsers.

10.4.4 Testing Plan

Different components of this application have different testing needs. This application will be used when people are out of their normal element, without knowledge of how things are normally done, and likely without a computer available. The directions, public transit maps, and 'when to leave' alerts in particular may be simple to design but glean useful information from field studies. This application should therefore use a mix of testing strategies to get the best information for the least investment of time and money.

As noted in Chapter 9, testing should be done with an emulator or actual device that shares the user interface paradigm of the device in the participant's pocket.

If the application were to be released as a corporate application to support corporate travelers, adding a usability component to the beta test, as described in the last section of Chapter 9, would provide excellent information for users like Justine, without the need for a more typical field test.

Appendix A

Mobile Markup Languages

WML 2 is based on WML 1.x and XHTML Mobile Profile, but it isn't widely implemented. XHTML Mobile Profile is based on XHTML Basic, and some browsers render it – mostly. XHTML Basic, with CSS, is widely implemented. Here's how all these technologies relate to each other.

The first markup language was SGML (Standard Generalized Markup Language), in 1974. It was good, but too complex. SGML was simplified – in both function and structure – to create HTML, a language-focused on presentation.

Later, the vision of a generalized markup language was rekindled, and XML (eXtensible Markup Language) was created as a (mostly) strict subset of SGML.

The Wireless Path

When Unwired Planet (later Phone.com, currently Openwave) wanted to create Internet access over a mobile phone, they analyzed factors such as device memory capabilities, wireless network connection and drop times, device display and control characteristics, and transfer speeds. They developed (although some will argue GeoWorks developed) HDML, or Handheld Device Markup Language.

Later, Openwave joined with Nokia and others to found the WAP Forum, now the Open Mobile Alliance. This group had the goal of creating a common standard for wireless Internet access. They largely took the features of HDML (with some exceptions that frustrated the usability community) and created WML (Wireless Markup

Language) as an XML language. This language proceeded from version 1.0 to 1.3, with version 1.1 the apparent most common browser implementation.

On the other side of the world, Japan's largest wireless carrier, NTT DoCoMo, created iMode (Information Mode) as a wireless service, running on their proprietary Compact HTML. This service became extremely popular, in no small part because of good price models and their restraint from marketing iMode as the web, but rather as information.

iMode and WML 1.x each have features not found in HTML. Few desktop users would find a special type of link to make a voice call to be particularly useful, yet it is critical on mobile phones. WML gave users access to commands associated with screens or items on the screen, not just hyperlinks. This allowed the scroll-and-select phones with one or two softkeys a bit more efficiency in accomplishing tasks on the phones.

One major problem with WML was the lack of standard rendering implementation. Some browsers rendered select lists as pop-up lists; other browsers rendered them inline (usually with no other components allowed on the screen). The result was that developers had to pick a browser to target and suffer an unacceptable user experience on the others, or double their work to target multiple browsers.

Figure A1 shows the relationships between various markup languages.

The W3C Path

Meanwhile, the W3C (World Wide Web Consortium) recast HTML 4 into XHTML 1, using syntactical rules from XML but the feature set (and tag names) of HTML. They then modularized XHTML into several units.

The W3C selected a set of modules appropriate for access by devices with limited capabilities. These included Basic Forms, Hypertext, and Basic Tables. They called this set of modules XHTML Basic.

One key advantage that XHTML Basic had over the other wireless markup languages was cascading style sheets (CSS) – useful now that most phones have graphical displays.

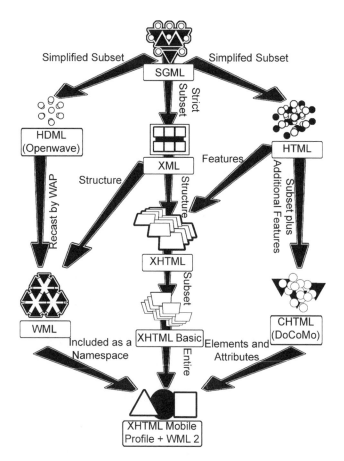

Figure A1 Relationships between various markup languages

The Convergence, Almost

NTT DoCoMo and the WAP Forum joined forces to create the next standard for wireless Internet access. They wanted to combine the features of WML, XHTML Basic, and iMode to create a platform that would serve all their users and developers.

They started with XHTML Basic to accelerate the convergence of wireless and desktop Internet development. With this, they got CSS.

They added in the functions from cHTML and WML that were not in XHTML Basic (but were in XHTML): acronym, address, br, b, big, hr, i, small, dl, fieldset, optgroup. The resulting language is a superset of XHTML Basic, but a subset of XHTML. They called this language XHTML Mobile Profile.

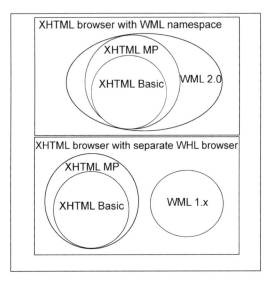

Figure A2   Implementation choices for WML 2 browsers

They then added the features of WML that could not be found in XHTML. These included navigation aids, onenter events, contexts, and other features (both elements and attributes). In true XML fashion, these were placed in an XML namespace and could be used by putting 'wml:' in front of the command. The combined XHTML Mobile Profile plus WML namespace is WML 2 (see Figure A2).

This, as far as usability is concerned, was the best possible solution. CSS gave control over fonts and layout. Designers could control cache, assist navigation, and have multiple non-link commands on a page using the WML namespace.

Enter politics. The wrong person made an off-the-cuff remark in front of the wrong people, and some of the Alliance members who wanted the mobile web to be just like the desktop web took the comment to heart. The Open Mobile Alliance, led by Nokia and NTT DoCoMo, determined that the WML namespace was 'just for backward compatibility'. Once this decision was made, the one to make the WML namespace optional quickly followed.

The Open Mobile Alliance decided that since the WML features were for backward compatibility, then a device could be WML 2 compliant either if it read WML 2, or if it read XHTML Mobile Profile pages and WML 1.x decks. There was no need to be able to read WML tags in the XHTML document.

Current State of Affairs

Nokia immediately created an XHTML Mobile Profile browser, with no WML namespace. The Openwave uses the WML namespace, which is to be expected since most of the features date back to HDML.

Most other browsers – such as the Access Compact NetFront (successor to the iMode browser) and the Samsung device browsers – support only XHTML Basic. Unfortunately, while these browsers generate very pretty pages, the usability will frequently be worse than the much-berated WAP.

Appendix B
Domain Names

Web sites have several options for providing user access to their mobile versions:

- separate domain names: www.mobileabc.com versus www.abc.com
- one domain name: www.abc.com, with no difference between mobile and desktop versions
- one domain name: www.abc.com, paired with device detection to deliver the mobile experience
- mobile subdomain: mobile.abc.com versus www.abc.com
- mobile top level domain: www.abc.mobi versus www.abc.com

Separate domain names will almost never be a good solution. They diminish the brand, add user confusion, and are not predictable. The use of a single domain without device detection should only be done with least-common-denominator designs – that is, rarely. Thus one of the last three options should normally be chosen.

Use of a single domain name coupled with device detection is easy for the user to remember, reduces information that must be delivered in advertisements, and is expandable to all types of potential access, such as interactive televisions. Its chief drawback is that users may not be confident that the site works well on the mobile.

A mobile subdomain can be used to give users the confidence that the site is targeted at mobile, as well as the flexibility to choose whether they want the mobile or the full version. This can be a good solution if the full version has many more features than the mobile version that are infrequently used *and* users are sophisticated in browser use. It

Designing the Mobile User Experience   Barbara Ballard
© 2007 John Wiley & Sons, Ltd

is an acceptable solution when promoting the mobile version of the site, encouraging users to try mobile. Tip: instead of 'mobile.abc.com' choose 'm.abc.com' as it requires less typing.

A mobile top-level domain, .mobi, has the same strengths and weaknesses as a mobile subdomain, with the extra cost of having to pay for the second domain name and fulfill their requirements process. Theoretically fulfilling these requirements gives users confidence that the site will work well on the mobile. Consider surveying your users as to whether this is true for them. Regardless, follow good design practices, beyond the W3C and .mobi requirements.

The question of domain names is most important when trying to get users to enter a URL, as when they are already browsing, clicking a link is a low-cost action. Advertising copy tends to have a browsing call to action: 'Visit us on the web at www.abc.com'. Any version of multiple domain names, either subdomains, .mobi, or simply dual domains, will require 'Visit us on the web from your PC at www.abc.com, or from your phone at www.abc.mobi', else the user will not know that there is a second domain. To communicate that a single domain name will work well on both, just shift the original copy a little bit: 'Visit us from your phone or PC at www.abc.com'.

Appendix C
Minimum Object Resolution

What is the smallest object that can be seen in a given image? The answer drives decision-making on composition of shots, and is not something that most photographers have considered.

Professional equipment is aimed at generating content for higher resolution display devices such as standard television. Camera professionals have an intuitive sense of how large an object has to be on the screen to work well for the viewing audience; this sense will not necessarily provide a satisfactory small-screen consumption experience. This appendix provides easy approximations for estimating whether or not a given camera shot has a chance of being usable on a mobile device.

Note that this is not a problem when authoring content using the same device as the target audience, since what the photographer's screen displays reflects what the viewer will see.

The equations below presume digital image capturing, thus eliminating such traditional technical film photography factors as film grain size, chromatic aberration. Pixels, both in capture and display, are the limiting factor.

For the minimum resolution for a target device we need to know the following:

- the angle of view of the lens creating the image, AoV
- the distance from the lens to the subject, D
- the pixel dimensions of the target display, the narrower of PixelsWide and PixelsHigh
- angle of view from camera specification.

Designing the Mobile User Experience   Barbara Ballard
© 2007 John Wiley & Sons, Ltd

Angle of view of a lens and sensor combination may be calculated using the following formula:

$$AoV = 2^*arctan\,\{(SensorDimension\,/\,[2^*FocalLength$$
$$^*(1 + Magnification)]\}$$

Sensor dimension, SD, is the diagonal measure of the camera sensor, or film for a film camera. Focal length, FL, is the reported size of the lens. The magnification factor is irrelevant except in macro- or micro-photography, eliminating that part of the equation. We are left with:

$$AoV = 2^*arctan\,\{SD/[2^*FL]\}$$

Take a 50-mm lens and put it on a Nikon D100 digital SLR with a 28.31-mm sensor, and the $AoV = 2^*arctan\,\{28.31/2^*50\}$, or 31.61 degrees.

Angle of View from Experimentation

You can also determine the angle of view for a given lens or given focal length setting on a zoom through practical experimentation.

Take an object of known length, and set it up so that its length is perpendicular to the axis of the lens. Move the lens until the object exactly fills the viewfinder width. Measure the distance from the object to the camera.

If the object is truly something small, like a yardstick, and the lens is a wide-angle lens, make your best estimate as to where the sensor is on the camera, as missing by a few centimeters will make a real difference at a hundred meters. You should be able to get within two centimeters, which should be good enough.

With the measured distance to and size of the target, the angle of view is:

$$AoV = 2^*arctan\,\{TargetSize/[2^*DistanceToTarget]\}$$

Minimum Viewable Object

The minimum viewable object at a given distance is determined by the smallest size that will map onto a single pixel, or:

$$\text{Minimum Viewable Object} = \text{Field of View}/\text{PixelsWide}$$

The total field of view for a given distance is:

$$\text{Field of View} = 2^*D^*\tan\{AoV/2\}$$

Thus for a typical sports scenario using the camera and lens discussed earlier, the field of view at 60 m distance is almost 34 m wide. At that distance, the minimum viewable object for a 240-pixel wide device is 34 m/240 = 14 cm. Not only will the ball disappear from view, so will the players.

Appendix D
Opt-In and Opt-Out

Marketing campaigns frequently use messaging services, including premium SMS, for simple, inexpensive, and effective communications. These campaigns can cost the user. The practices in this appendix summarize methods for ensuring the user has agreed to the additional charges.

Opt In

A single opt-in is used for simple program participation, at no charge beyond messaging fees. Note that in the United States, many users pay for each message received. In regions where this is true for programs with high volume, consider a double opt-in.

User experience

- User sends text message to the service short code.
- Server responds with program, opt-out, and contact information, including voice number if possible.

The program, opt-out, and contact information should be resent when the user sends 'help' to the short code. A URL for a website can be included if it will provide superior customer service.

Designing the Mobile User Experience   Barbara Ballard
© 2007 John Wiley & Sons, Ltd

The company should:

- avoid sending user messages for any other program without specific permission
- avoid selling user contact information without specific permission.

Double Opt In

A double opt-in is used for premium SMS or other situations when the user will be charged. This is an expansion of the opt-in recommendations.

User experience, first-time sending to a short code for a new service

- User sends text message to short code.
- Server responds with program, opt-out, and contact information, including voice number.
- Server responds with pricing terms, including amount of money, frequency, and that it will be deducted from month bill or pre-pay balance.
- Server sends request for participation verification.
- User sends affirmative response of any flavor (yes, oui, sure, etc.)
- Server sends requested content and charges user.

The first response can include a URL and PIN for interaction via the Internet, either desktop or mobile.

User experience, subsequent uses

- User sends text message to the short code.
- Server sends requested content.

The program, opt-out, and contact information should be resent when the user sends 'help' to the short code. A URL for a website can be included if it will provide superior customer service.

The company should:

- avoid sending user messages for any other program without specific permission

- avoid selling user contact information without specific permission
- provide periodic opt-out information.

Opt Out

Services should advise users to use 'stop' to stop receiving messages at the beginning of all programs, throughout pay programs, and on advertising copy. Further, the server needs to respond to ill-formed opt-out requests at any time.

If the user sends 'end', 'stop', 'cancel', 'unsubscribe', 'quit', or any cognates like 'go away' to a short code, the service should stop. If any of these is followed by 'all', 'anything', or cognates, all services provided to that server from that short code should be stopped. As with all SMS interaction, case should be ignored.

If multiple services are provided by that short code to that user, stop the most recently used service – unless the user has named a program after the stop command.

The server should send a free confirmation message, indicating that the service has been canceled.

Appendix E
Mobile Companies

Companies enter and exit the mobile industry, making any attempt at a comprehensive list sure to be inaccurate as soon as it is compiled. Regardless, many companies have ongoing impact on the industry. This appendix includes major industry influencers, a sampling of second-tier players, and a cross-section of industry company types, from carriers to content providers. See the Chapter 8 for an overview of the company types and roles.

Access

Mobile software company. Its browsers are in many handsets; it purchased Palm OS with a focus on shifting it to Linux. It believes Linux is the best OS for the Asian market.
Main site http://www.access.co.jp/english
Developer support http://www.palmsource.com/developers

Adobe

Along with many other things well known to most design professionals, Adobe's Macromedia unit is the developer of the Flash Lite platform.
http://www.adobe.com/devnet/devices/flashlite.html

AOL

Internet access provider turned Internet services provider. Entered mobile space in 1999 with email and similar applications. purchased Tegic and then made the mobile text entry company be the AOL mobile business unit.
http://www.aol.com

China Mobile

Dominant operator in China, largest mobile operator in world by number of subscribers; uses GSM.
http://www.chinamobile.com/ENGLISH/index.html

China Unicom

Major operator in China; uses CDMA.
http://www.chinaunicom.com [Chinese only]

Cingular

One of top two US carriers, depending on the year and whether any recent acquisitions were made. GSM carrier.
Main site http://www.cingular.com
Developer support http://developer.cingular.com

Danger

Manufacturer of the Danger Sidekick consumer messaging device
Main site http://danger.com
Developer support http://developer.danger.com

EA (Electronic Arts)

Electronic Arts, the game developer and distributor, purchased Jamdat Mobile, perhaps the top Java and BREW game developer. The

combination has EA as the dominant player in the mobile game market. Expect competition from Digital Chocolate
http://www.eamobile.com

Garmin

Dominant GPS device maker, with Palm and PocketPC devices as well
http://www.garmin.com

Jamdat

Game developer purchased by EA. See EA.

KDDI

Japanese carrier
http://www.au.kddi.com/english/index.html

LG

Korean phone manufacturer, with a history of working closely with operators to deliver phones to the operator's specifications
http://www.lge.com

Lucent

Infrastructure and technology provider, including switches and location technology
http://www.lucent.com

Microsoft

Developer of software and Windows Mobile Smartphone and PocketPC editions
Main site http://www.microsoft.com
   Mobile developer support http://msdn.microsoft.com/mobility

MobiTV

Mobile television provider
http://www.mobitv.com

Motorola

Device manufacturer and infrastructure provider, diffentiating on industrial design
Main site http://www.motorola.com
Developer support http://developer.motorola.com

Motricity

Content aggregator and application storefront
http://www.motricity.com

Nokia

Largest phone manufacturer, differentiating on design
Main site http://nokia.com/index.html
Developer and designer support http://forum.nokia.com

NTT DoCoMo

Largest Japanese wireless carrier, inventor of the popular and profitable iMode wireless data services
Main sites http://www.nttdocomo.co.jp/english/index.html and http://www.nttdocomo.com
Developer support http://www.nttdocomo.co.jp/english/p_s/i/make/index.html and http://www.doja-developer.net

Orange

Mostly European mobile operator with British history and French (France Télécom) ownership
Main site http://www.orange.com
Developer support http://www.orangepartner.com

Palm

Maker of handheld electronic organizers. Palm split into Palm and PalmSource, then PalmSource sold themselves to Access of Japan. Expect Palm to migrate towards a different operating system; some of their devices already run Windows Mobile
Main site http://www.palm.com
Developer support http://www.palmsource.com/developers [owned by Access]

Qualcomm

Original developer of CDMA technology; continues to license the technology to device manufacturers. Former device manufacturer, but sold the division to Kyocera. Developer of the BREW platform, which was originally intended just to help the company quickly develop handsets. Purchased Trigenix, inventors of TrigML and related technologies, and renamed the technology uiOne and integrated it with BREW
Main site http://www.qualcomm.com
Developer support http://brew.qualcomm.com [click on Developer]

Reliance Infocomm

Dominant Indian operator; uses CDMA
Main site http://www.relianceinfo.com
Developer support http://www.dadp.com

Research in Motion

Manufacturer of BlackBerry business messaging devices
Main site http://www.rim.com
Developer support http://www.blackberry.com/developers

Samsung

Korean phone manufacturer, consistently in the top five in sales volumes. Standard strategy is to adopt any operating system or technology that comes available
http://samsung.com/products/wirelessphones/index.htm

SK Telecom

Dominant Korean carrier, expanding internationally
http://www.sktelecom.com/eng

Sony Ericsson

Joint venture of Sony and Ericsson, with access to Sony's entertainment technologies and Sony BMG content. Major European device manufacturer
Main site http://www.sonyericsson.com
Developer support http://developer.sonyericsson.com

Sprint Nextel

Sprint was the number three US carrier when it merged with business-centric Nextel. CDMA carrier with a history of influencing device design. Pure wireless carrier
Main site http://www.sprint.com
Developer and designer support http://developer.sprint.com

Sun

Owner and coordinator of the Java platform, including Java ME
Mobile developer support http://java.sun.com/javame

Symbian

A consortium of device manufacturers, dominated by Nokia, and the operating system they jointly developed. Different platforms with different user interfaces run Symbian, notably the Nokia Series 60, Nokia Series 80, and UIQ platforms
Main site and developer support http://www.symbian.com

Symbol Technologies

Maker of highly rugged scanner devices, including some PocketPC and
Palm units
http://www.symbol.com

T-Mobile

A multi-national GSM carrier, owned by Deutsche Telekom. Note that
a USA customer visiting the UK will still have to pay roaming charges
http://www.t-mobile-international.com

Tata Indicom

Dominant Indian operator; uses CDMA
Main site http://www.tataindicom.com

Tegic

Makers of letter prediction software Tegic T9. Was purchased by AOL
and now forms the core of their mobile unit.
http://www.tegic.com

Telus Mobility

Innovative Canadian wireless operator
http://www.telusmobility.com

Verizon

One of top two US carriers, depending on the year and whether any
recent acquisitions were made: 45 % ownership by Vodafone; a CDMA
carrier
Main site http://www.verizonwireless.com
Developer support http://www.vzwdevelopers.com/aims

Visto

Push email platform provider predating the BlackBerry
http://www.visto.com

Vodafone

European carrier with large international presence, owning parts of
carriers in countries where they have no presence
  Main site http://www.vodafone.com
  Developer support http://www.via.vodafone.com

Zi Corporation

Developers of letter and word prediction software like eZiText
  http://www.zicorp.com

Glossary

1xRTT An intermediate standard between 2G (less than twenty kilobaud connection) and 3G for CDMA carriers. Key characteristics include packet data, the ability to give priority to some connections over others, easy upgrade from 2G, and speeds up to 150 kilobaud. Average speeds are approximately 50 kb/s.

2 G Digital wireless voice communications. If data transmission is possible, it is limited to less than 20 kb/s. GSM, CDMA, iDEN, PDC, and TDMA are all 2G technologies.

2.5 G Digital wireless communications. Data transmission is always packetized rather than a dial-up connection. Data speeds average around 50 kb/second and can burst up to 200 kb/s so. Acronyms like GPRS, CDMA EDGE, and 1xRTT abound.

3 G Digital wireless communications, becoming common in 2006. Acronyms like EVDO (CDMA-based) and W-CDMA (GSM based). The Chinese have their own version, TD-SCDMA. Video messaging and calls are possible with data speeds up to 2 Mb/s.

3GPP (3rd Generation Partnership Project) Consortium providing specifications for GSM-based 3G technologies. Of particular note are media standards like MPEG-4 for mobile.

3GPP2 Consortium providing specifications for CDMA-based 3G technologies. Of particular note are media standards like MPEG-4 for mobile.

4 G Digital wireless communications, purportedly with broadband speeds. WiMAX is an example technology.

adaptive design The practice of designing one application that detects device capabilities and alters rendering based on this information. One common technique is using multiple CSS files for the same web site. Can work well for simple applications whose structure does not need to change with significant changes in device capabilities. Results in a fairly good user experience, but at odds with the native user interface style of the device. See also *class-based design, least-common-denominator design*, and *device targeted design*.

AJAX Asynchronous JavaScript and XML. A collection of technologies which combine to give browsers more immediate interaction with the user, similar to fully fledged application languages. Support for these technologies varies wildly. See also *ECMAScript*.

automatic translation The use of database engines, application logic, device characteristics, and simplifying assumptions to render applications on multiple devices natively. This approach can result in good applications if its scope is limited by either a highly targeted audience or a focus exclusively on PCDs with one input mechanism (scroll and select vs. stylus).

BlackBerry A PCD, made by Research in Motion, focused on email and text communications, very popular with the corporate and blind communities. Sometimes colloquially called a 'CrackBerry', indicating the degree of addiction some users feel for it.

Bluetooth Wireless communications protocol used for local communications, with a range limited to a person's immediate vicinity. Common uses include wireless headsets for voice communications and synchronization. This is one useful method for enabling a pico net.

breadcrumbs On web sites, the practice of indicating where in the site hierarchy the current page sits by providing a set of links, typically in a small font, to each of the current page's parents. A typical design might be Home > Resources > Design documents > Current page. See also *signposting*.

BREW An application environment with significant advantage on CDMA phones, as the environment grew out of Qualcomm's device coding environment. The platform has been extended to GSM phones. The language is based on C++. Deploying a BREW application requires carrier involvement.

calling party pays The European practice of charging the calling party for mobile termination of a call. Applies to SMS as well. See also *receiving party pays*.

carrier A synonym for operator.

cHTML (compact HTML) The reduced version of HTML used by NTT DoCoMo as the markup language for the iMode service. See also *iMode*.

class-based design The practice of designing for a set of classes, or collections of devices with similar capabilities and user interface styles. Classes can be quite general, such as 'scroll-and-select device', or quite specific, such as 'Nokia Series 60 devices' or 'Motorola RAZRs'. Costs slightly more during development, as the design effort is focused on a handful of classes rather than one, but with a user experience close to the native device user interface style. See also *device targeted design, least-common-denominator design*, and *adaptive design*.

connectivity   Any of a large number of methods a device can use to access remote data, including Bluetooth, Wi-Fi, GSM, GPRS, and so forth.

data plan   A typically add-on service from a carrier enabling a PCD to get data services in addition to voice services. Setting up a data plan with a carrier is not always simple for the user to accomplish, and represents a major barrier to use for many applications. Text messages are usually not counted as part of a data plan.

deck   A set of mobile web pages run by a single organization. Most commonly used referring to the 'carrier's deck', or the set of pages available when the user launches the browser. The term 'deck' derives from HDML and WML 1.x, in which each request would return a 'deck' of one to four related 'cards'; the term was built into the language.

design pattern   In software engineering, a common software design situation and a standard solution. The design situation may occur in different platforms and languages. A pattern may include behaviors, intent, consequences, known uses, and sample code – but not executable code. See also *user interface design pattern*.

device description repository   A list of devices and their various capabilities, to be used by an application to adapt itself to a specific device environment. Rarely if ever do these include user interface style. See also *user interface style, rendering engine, device hierarchy, WURFL*, and *J2ME Polish*.

device hierarchy   An organization of devices based on the user interaction characteristics that affect interaction design, like stylus input, softkey paradigm, and features and capabilities. This organization can be built by a development organization, and user interface design patterns are built based on nodes in the hierarchy. The input to the hierarchy starts with a device description repository. Contrast with device taxonomy, discussed in Chapter 3, which categorizes market segments for devices. See also *user interface design pattern, device description repository*, and *user interface style*.

device proliferation   Mobile devices come in a number of shapes and sizes, with varying features. Devices may focus on gaming, blogging, messaging, email, or voice. Each of these different device types, plus the manufacturers' need for differentiation, leads to a wide variety of input mechanisms, user interfaces, and rendering issues. If devices ever become 'standardized', they will be standardized in clusters; expect proliferation to continue.

device targeted design   The practice of designing for a specific small set of devices. Results in a highly optimal user experience on the targeted device, but at significant cost in either market penetration, development cost, or both. See also *class-based design, device hierarchy, least-common-denominator design*, and *adaptive design.*

ECMAScript   The language of the ECMA-262 specification; it is a vendor-neutral standardization of JavaScript. Support for ECMAScript is necessary before AJAX levels of interaction are possible. Some browsers support ECMAScript Mobile Profile, with the primary limitation of not supporting eval. See also *AJAX.*

electronic paper   A display technology that emits no light, instead relying on reflected light. It has many of the characteristics of paper with ink that can move. Only uses power when changing the display, but can only change the display a couple times per second.

electrowetting   A display technology that uses an electric field to decide whether a colored oil covers or doesn't cover the substrate. These displays have excellent color and low power consumption. Unlike electronic paper displays, they can also be changed at video speeds.

emulator   Software that uses the same rendering code as a specific mobile device or application, but displayed on a computer. These are useful for developing and testing code, and are quite reliable but rare. See also *simulator.*

ethnographic research   A research technique inherited from anthropology that involves observing potential users' entire context, patterns, practices, and needs associated with some type of concept. A possible mobile ethnography project would be investigating how people interact with and share music, both at home and away, with an eye towards creating a mobile device or application that enhances the experience and fits the needs. Ethnography is particularly good for creating brand new products, as users cannot yet articulate their needs and context for a nonexistent product.

Fastap   Digit Wireless' full alphabetic keyboard in the same physical space as a more traditional numeric keypad. This is accomplished by laying out numbers as usual, and then in the corners between the numbers putting letter buttons. The numbers themselves aren't really buttons, but are activated when three or four of the surrounding letter-buttons are pressed when the user 'presses' the number.

Fitt's Law   From ergonomics, Paul Fitt's model of target acquisition: the amount of time required to move (a hand, cursor, or pointer) to a target is a function of the size of the target and its distance from the current location. The larger the target and the closer the target, the faster the acquisition. See Chapter 5 for a discussion of applying Fitt's Law to mobile design.

Flash Lite   A combination of scalable vector graphics (proprietary) and scripting, which together make an application environment. Owned by Adobe (formerly Macromedia). Flash Lite has fewer scripting capabilities than full Flash. See also *SVG*.

geotagging   The recreational practice of adding location (typically latitude and longitude) data to online information, enabling local searches, nearby searches, and other services.

global positioning system (GPS)   A set of satellites broadcasting weak signals. GPS-enabled devices determine which satellites are visible, what the variances are in time stamps, and hence which satellites are further away. This combination allows the device to determine where it is within 5 meters, or 100 meters, or not at all – depending on conditions. Assisted GPS 'boosts' the effective signal by using the known location of the cell tower as part of the calculations.

gossip   A social behavior in which participants discuss characteristics and situations of people, either public or personally known, who are generally not present. Facilitates social grooming and community building.

HDML   Handheld Device Markup Language, the markup language developed by Openwave (then known as Unwired Planet) for delivery of text and simple graphic information to mobile phones. HDML was a major inspiration for WML 1.0. See also *WAP, WML,* and *XHTML Basic.*

high-fidelity (usability) testing   Usability testing with a prototype very similar to the final product in form, function, features, and visual design. Contrast with low-fidelity testing.

hiptop   A term coined by Danger to describe their Sidekick device, a play on 'laptop' and 'device worn on the hip'. See also *Sidekick.*

iMode   NTT DoCoMo's mobile Internet service system. See also *keitai, cHTML,* and *iMode ecosystem.*

iMode ecosystem   More than just a markup language, the entire business process for delivering iMode applications. Includes NTT DoCoMo's tight integration with device design and development, the markup language cHTML, the model for sharing revenues with developers, and the semi-walled garden with access to services outside DoCoMo's recommended services. See also *walled garden, deck,* and *iMode.*

information appliance   A computing device focused on accomplishing one task very well, to the exclusion of other tasks. Contrast with general-purpose computing device.

interoperability   The degree to which services like web sites and text messaging can work when shared by users with different types of devices or different carriers. A lack of SMS interoperability stifled the US market for text messaging.

J2ME Polish   An open-source build environment for Java ME that allows the designer to use high-level widgets but control their font, color, spacing, and layout using CSS. The output is an application with many versions but low testing. Includes a device description repository. See also *rendering engine, WURFL,* and *device description repository.*

Java Mobile Edition (Java ME)   Formerly J2ME, this is a collection of objects and classes written in Java. Objects familiar to fully fledged Java developers may not be present.

keiretsu   Japanese term for a cluster of companies with significant, deep, decades-long interconnections. Companies within a keiretsu tend to do business only with others within the keiretsu. Strong and long-lasting supplier relationships enable companies to adjust design and manufacturing processes for higher quality. Korea adopted a similar but family-centered practice, chaebol, with large cross-industry conglomerates like LG, Samsung, SK Group, and Hyundai.

ketai   Japanese term for the mobile phone, particularly the mobile phone as an Internet device. See also *iMode.*

KiloByte Virtual Machine (KVM)   The software, resident on the mobile handset, that enables a Java ME application to run. In computing terms, the Java interpreter. See also *MIDP.*

landline   A telephone connection using direct copper from point to point. Contrast to wireless and even voice-over-IP. All telephones prior to roughly 1985 were landline phones.

least-common-denominator design   The practice of designing one application and one user interface for every conceivable device that will use the application. Results in a suboptimal experience on every device. See also *class-based design, device hierarchy, device targeted design,* and *adaptive design.*

Likert scale   Numbered responses with clear labels (such as Strongly Agree, Agree, Neither Agree Nor Disagree, Disagree, Strongly Disagree) in a questionnaire. Contrast with semantic differential scales, which do not use labels.

location-based services   Applications that use the location of the device, as determined by GPS, tower location, proximity of a specific wi-fi device, or other methods. See also *global positioning system.*

low-fidelity (usability) testing   Usability testing of a simulation of the final product, with key aspects unlike the tested design. Most common application is Wizard of Oz testing, in which participants use a hand-drawn paper prototype rather than a coded system. See also *high-fidelity testing.*

MIDP (Mobile Information Device Profile)   When combined with the Connected Limited Device Configuration, provides by far the most common definition, programming interface, and environment for running Java Mobile Edition applications. Note that a large number of APIs are considered optional, so one MIDP device does not necessarily behave the same as another. See also *KVM* and *Java ME*.

MMS (Multimedia Message Service) Similar to SMS, but allows pictures, text, and sound to be transmitted to and from a mobile. See also *SMS*.

mobile A device or service used by a user who has the potential to move to a new location, even a new building or city, during use. Includes both automobile and phones. This book focuses on handheld communications. See also *personal communications device*.

Mobile Virtual Network Operator (MVNO) A wireless operator that does not manage its own towers or network operations, instead buying connectivity from a traditional mobile operator.

mobilizing The process of converting a desktop application's features, navigation, design, and even content to match the needs of mobile users and devices.

network A service, accessible from a large area, provided by an operator providing wireless access to remote data, including voice and data connectivity.

Nielsen, Jakob Well-known usability guru.

Nokia-style softkeys From the Nokia standard user interface, the software labeling of two softkeys as 'Options' and 'Cancel' (or 'Back' or 'Exit', depending on circumstances). Options brings up a menu of available commands. See also *softkeys*.

OLED Organic Light-Emitting Diode, a display technology that uses organic compounds to generate light. Uses less power than traditional LCDs.

operator The organization providing the connection between the wireless device and both the Internet and other remote voice devices. A synonym of 'carrier'. Major carriers include NTT DoCoMo, Orange, Deutsch Telekom, and Verizon.

OS (operating system) The core software that provides a device's user interface paradigm and gives all programs a context within which to run. Some operating systems have publicly known names: Linux, Symbian, Windows Mobile, Palm OS. Others exist, but are used only by the device manufacturer. Frequently 'operating system' will refer only to the first type.

page The core displayable unit, particularly an XHTML page. The amount of information on a page can exceed the information visible on the screen, and is accessible by scrolling. See also *screen*.

PDA (personal data assistant) A term that varies throughout the industry. The most common understanding is a mobile device with no voice connectivity and the ability to run downloaded applications. Using this definition, PDAs represent a disappearing market segment (or a rapidly expanding one if you include certain GPS devices). See also *smart phone.*

persona In the design process, a personification of a collection of goals, behaviors, and context gathered from the user research process. Personas were popularized by Alan Cooper in *About Face.*

personal communications device (PCD) A handheld device focusing on communications, voice or text or both. Includes Danger Sidekick, RIM Black-Berry, and most mobile phones. Key characteristics are that the device is personal, communications-focused, wakable, and handheld. See also *The Carry Principle* for resulting characteristics.

pico net A network of devices focused on a single person. Can include music players, mobile phones, other personal communications devices, wireless head-sets, and even personal computers.

platform An application technology and an environment in which it runs. It includes, at a minimum, a development language and an interpreter or compiler loadable on a device. Examples include web (XHTML and browser) and Java ME (Java MIDP and KVM).

portal A collection of web pages intended to be a primary access to the Internet. Examples include Yahoo!, AOL, and carrier home decks.

porting The practice of converting an application from one platform to another or from one device to another.

post-paid A type of subscriber agreement with a carrier. The subscriber receives credit for each month's service, and pays the bill at the end of the month; typically a minimum monthly charge applies. The need for credit worthiness means that populations with insufficient credit access cannot use this method; these populations include teenagers and much of Africa. The need for significant personal data means that people who wish extra privacy cannot use this method. See also *pre-paid.*

pre-fetch data An application can wait until the user requests information to fetch it to the device. This reduces network traffic and potentially user data charges, but reduces application responsiveness. Alternately, the application can make smart predictions about what the user will need next, like specific graphics or the shell of a page without specific data yet loaded, and fetch that from the network before the user requests it. This increases perceived application speed.

pre-paid A type of subscriber agreement with a carrier. The subscriber pays for a phone and some number of minutes of use. The cost per minute tends to be higher than post-paid agreements and churn can be quite high without obligation to stay with a carrier. See also *post-paid.*

premium SMS   SMS message costing extra money, used as a mechanism to pay for content or services. See also *SMS* and *short codes*.

QVGA   Quarter Video Graphics Array. A standard screen dimension, 320 by 240 pixels. For many phones, the vertical dimension is larger. Note that mobile devices have grown to be this large, but a quarter this size (160 by 120) is more common, and desktop computers have long since left full VGA (640 by 480) behind.

QWERTY keyboard   Mobile devices can have a QWERTY layout for the keyboard, with one button for each letter. Supplemental buttons including Shift, Tab, Return, and numbers are rarely in the same location as on a computer keyboard. Mobile QWERTY keyboards must be small enough to be operable with two thumbs, so full hand touch-typing is not possible.

receiving party pays   The largely American practice of charging the mobile subscriber for receiving a call. The calling party also has to pay for network access, but at lower rates than in calling party pays markets. Applies to SMS as well. Increases mobile subscriber concerns with spam, telemarketing, and receiving calls in general. See also *calling party pays*.

rendering engine   A typically database-driven piece of software that attempts device capabilities detection and then reformatting or even restructuring presentation, optimizing for specific device capabilities. See also *device description repository, user interface style, WURFL*, and *J2ME Polish*.

rendering idiosyncrasies   How an application is rendered depends on the application code and the platform, but also how the application environment company understands the platform specification, device capabilities and user interface, and any design decisions the application environment company and the device manufacturer made. As a result, parts of an application may be broken and other parts may not display as expected. While some of the resulting idiosyncrasies could be eliminated through quality control processes across the industry, others are a function of necessary device differences.

repurposing   The practice of taking existing content and converting it for use on the mobile phone. When done well, the mobile user experience is enhanced. When done poorly, the content merely seems to be a lower quality fewer feature version of what is available on the desktop. See also *mobilizing*.

RFID (Radio Frequency Identification)   A type of chip that can be embedded in devices, or even dogs, and can be read nearby. Depending on the context, the information can be used to identify a device for use as a mobile wallet – the user just waves the phone over the reader at the point of sale, and the transaction is recorded.

screen   (a) The physical display on a device. (b) In an application, the information that is visible at one time. Page-based applications like most web sites and some information-driven local applications can have information beyond the confines of the screen, accessible by scrolling. Screen-based applications provide no information visible beyond the screen, and have no page scrolling necessary. See also *page*.

scroll-and-select   A device whose selection cursor is moved by scrolling, usually a button on a two- or four-way navigation control, until the designed control or item is highlighted. A separate button selects and activates the highlighted item. See also *stylus device*.

Section 255 of the Telecommunications Act of 1996   United States regulation requiring telecommunications products to be accessible by people with various disabilities.

Section 508 of the Rehabilitation Amendment Act of 1998   United States regulation requiring telecommunications and computer equipment and software purchased by the government to have best-of-class usability by people with various disabilities.

semantic differential scale   In a questionnaire, a response with usually five, seven, or nine ticks, with only the end points labeled. End labels are opposite: Agree and Disagree or another pair. Contrast with *Likert scales*, which have specific labels for each point on the scale.

short code   Special phone number, owned by a content provider, of five or six digits for texting. See also *SMS* and *premium SMS*.

Sidekick   A PCD focused on instant messaging and email, made by Danger. Very popular with teens. See also *hiptop*.

signposting   The practice of providing visual information on a screen indicating where in the application the user currently is. See also *breadcrumbs*.

SIM (Subscriber Identity Module)   For GSM phones, the information that allows a phone to function by identifying the carrier and subscriber associated with the phone.

simulator   Software that simulates the mobile experience, but displayed on a computer. Because the code used to display information is not the code used for the mobile, expect rendering and other implementation differences. Simulators can be used only for first-pass component-level testing; they cannot be used for system or user testing. See also *emulator*.

smart phone   A term that varies throughout the industry. The most common understanding is a mobile phone with a named operating system that allows downloaded applications, particularly Symbian, Palm, and similar devices. Whether this includes devices with closed operating systems that support Java and BREW varies. Note also that Microsoft has named the scroll-and-select version of their Windows Mobile as SmartPhone Edition, which makes many people refer to the devices as smartphones. There have been erroneous reports that Microsoft coined the term. See also *PDA*.

SMS (Short Message Service)   Text messages of up to approximately 160 characters sent to or from a mobile. Entire applications can be written using SMS as the display and communications mechanism. See also *MMS, premium SMS,* and *short codes.*

softkey   A hardware button without pre-printed label, and the associated variable function labeled on the screen next to the button. Functions associated with the button can be specific to the highlighted item, or general to the screen. Most scroll-and-select devices use softkeys to provide functionality beyond simple selection of an item. Stylus-driven devices typically use pure software buttons or menus to accomplish the same goal. See also *scroll-and-select* and *Nokia-style softkeys.*

stylus device   A mobile device operated primarily with a device using a touch screen. See also *scroll-and-select* and *PDA.*

SVG (Scalable Vector Graphics)   An open-standard graphics and interactive application delivery mechanism, similar to Adobe/Macromedia Flash. Contrast with bitmaps, which have to be manually resized. See also *Flash Lite.*

Symbian   A consortium of device manufacturers, dominated by Nokia, and the operating system they jointly developed. Different platforms with different user interfaces run Symbian, notably the Nokia Series 60, Nokia Series 80, and UIQ platforms.

Tegic T9   Letter prediction software for facilitating text entry on a standard keypad. To type a word, the user presses each letter button once; the software looks in its dictionary to see what words match the 3 * *n* (number of letters) combinations of what the user has entered. If more than one word matches, the user chooses which one is intended. See also *triple tap* and *WordLogic.*

telco   Telecommunications company, usually landline or wireless operator.

texting   A shorthand for 'communicating using SMS messages'. See also *SMS.*

The Carry Principle   The idea that because a mobile is always carried, it will have a number of important characteristics: wireless communications, multi-function device, battery powered, small, personal, and always on. This in turn has a number of device and application design implications.

theme   A recasting of a device's native user interface with altered background graphics, fonts, standby screen, softkey images, visual design, and sometimes functionality.

thumb keyboard   A QWERTY or similar full-function keyboard scaled down such that all buttons are operable by the thumbs while the user is holding the device. Typical devices include the BlackBerry, Palm Treo, and Nokia Communicators.

triple tap   Base text entry mechanism for a 12-button keypad. To enter a 'C' on most English language phones, press the '2' button three times because C is the third letter on the button. All other text entry mechanisms compare themselves to this base. See also *Tegic, WordLogic*, and *Fastap*.

ubiquitous computing   Computing capabilities, and information sharing, in a wide variety of devices ranging from phones to chairs to coffee cups.

ubiquitous web   The presence of a Internet connection, with varying features, on a wide variety of devices ranging from phones to refrigerators to wall displays.

uiOne   A collection of technologies that allows operators and others to completely customize the software of a uiOne-enabled mobile phone. Functionality can be added or removed, the user interface paradigm can change, graphics and animations can change, labels can change. Perhaps most importantly, the standby screen can change. In practice, operators do not allow this full amount of customization, but 'themes' can change limited amounts of the user experience, enabling branding and personalization without risking core function. Owned by Qualcomm and integrated with BREW.

usability design pattern   A user interface pattern or development process that reliably results in software with good usability. See also *user interface design pattern*.

user interface design pattern   A user interface structure, including widgets, behaviors, and sometimes visual elements, that is a standard (and hopefully good) response to a common design pattern. See also *design pattern* and *device hierarchy*.

user interface style   A device operating system's input method, softkey use policy, information organization, select key use, and visual design. Both Nokia and Samsung phones tend to have two softkeys, but they are used quite diffrently. The Nokia device will in most situations display 'Options' on the left softkey; the Samsung device may have the left softkey serve as the select function, or may have it be selection dependent. A device's user interface style strongly influences its user's expectations. See also *Nokia-style softkeys* and *device hierarchy*.

voice-over-IP (VoIP)   The ability to transfer voice communications over computer networks. This goes beyond 'broadband telephone' (like Vonage) and creates the ability to create mixed visual and aural services, asynchronous voice communications like Voice SMS, and many others. See also *voice SMS*.

voice SMS   A short voice message delivered using SMS or MMS as a mechanism. Some devices receive the entire message and can play it locally; other devices receive a link to the content so there is a delay in retrieval. See also *voice-over-IP* and *SMS*.

VoiceXML   A markup language intended for voice input and aural output. The structure of the language is quite different from visual markup languages.

walled garden   The collection of web sites and applications accessible from the carrier's home deck, when the carrier blocks access to other services. Carriers can block access by removing the 'enter URL' function from the browser while simultaneously providing only search within the carrier's content.

WAP (Wireless Application Protocol)   A collection of technologies, including WML, XHTML MP, location services, WAP Push, session protocol, security layer, location services, and so forth. The standards are owned not by the W3C, but the Open Mobile Alliance. See also *WML, XHTML Basic, WAP Push*, and *XHTML Mobile Profile*.

WAP Push   A SMS message with special formatting and a URL pointing to a WAP site. The user receives the information and the ability to visit the site immediately. This can be used for application delivery as well as important content updates.

Wi-Fi   A wireless data transfer protocol over unlicensed bandwidth. Used particularly for Internet access.

wireframe   A representation of a site or application screen function and content layout. Explicitly ignores color and font choices.

WML (Wireless Markup Language)   A small markup language designed for mobile phones, based on HDML. Typically refers to WML 1.x, which was replaced by XHTML Mobile Profile and WML2. See also *WAP, HTML, XHTML Mobile Profile*, and *WML 2*.

WML 2   XHTML MP plus wml extensions (or wml namespace). Rarely used. See appendix *Mobile Markup Languages* for details. See also *HDML, WAP, WML, wml namespace*, and *XHTML Mobile Profile*.

wml namespace (wml extensions)   The collection of tags and attributes from WML 1.3 that were not anywhere in XHTML. Includes cache management features and navigation features. While the original intent was to include this as a required part of WML2, a compromise was reached wherein a WML2 browser would be considered compliant if it simply supported XHTML and WML 1.2 in separate pages. See also *WML, WML2, XHTML Basic*, and *XHTML MP*.

WordLogic   Word and letter prediction software that works on Windows Pocket PC, Tablet, and Desktop versions. On stylus devices, uses virtual keyboards, highlighting of possible and most likely next letters, word completion pop-ups, and other word entry optimizations.

WURFL (Wireless Universal Resource File)   An open-source device description repository. See also *J2ME Polish, rendering engine*, and *device description repository*.

XHTML Basic   The smallest subset of tags necessary to create a functional web site, formalized into an XHTML module. Mobile browsers support at least XHTML Basic. See appendix *Mobile Markup Languages* for details. See also *HDML, WAP, WML*, and *XHTML Mobile Profile*.

XHTML Mobile Profile (XHTML MP)   XHTML Basic with the addition of a few tags such as b. See appendix *Mobile Markup Languages* for details. See also *WML* and *XHTML Basic*.

Index
